The Newtonians and the English Revolution 1689–1720

Classics in the History and Philosophy of Science

Series Editor
Roger Hahn, University of California, Berkeley

This book is part of a series. The publisher will accept continuation orders which may be cancelled at any time and which provide for automatic billing and shipping of each title in the series upon publication. Please write for details.

The Newtonians and the English Revolution 1689–1720

by

MARGARET C. JACOB

New School for Social Research
New York, USA

Gordon and Breach

New York   Philadelphia   London   Paris   Montreux   Tokyo   Melbourne

© 1976 by Cornell University. Reprinted 1990 by Gordon and Breach Science Publishers S.A. by permission arrangements with Cornell University Press.

86967147

Gordon and Breach Science Publishers

Post Office Box 786
Cooper Station
New York, New York 10276
United States of America

5301 Tacony Street, Slot 330
Philadelphia, Pennsylvania 19137
United States of America

Post Office Box 197
London WC2E 9PX
United Kingdom

58, rue Lhomond
75005 Paris
France

Post Office Box 161
1820 Montreux 2
Switzerland

3-14-9, Okubo
Shinjuku-ku, Tokyo 169
Japan

Private Bag 8
Camberwell, Victoria 3124
Australia

BF
1623
.P5
235
1990 X

Library of Congress Cataloging-in-Publication Data

Jacob, Margaret C., 1943-
    The Newtonians and the English Revolution, 1689–1720 / Margaret C. Jacob.
        p.   cm. — (Classics in the history and philosophy of science ; 7)
    Reprint, with new pref. Originally published: Ithaca, N.Y. : Cornell University Press, 1976.
    Includes bibliographical references.
    ISBN 2-88124-400-9
    1. Religion and science—England—History.   2. Science—England—History. 3. Latitudinarianism (Church of England)—History.
    4. England—Intellectual life—17th century.   5. England—Intellectual life—18th century.   6. Great Britain—History—Revolution of 1688.   I. Title.   II. Series.
BL245.J3   1990
306.4'5'094209032—dc20   89-28361
                                                                                          CIP

To my mother
In memory of my father

Contents

Preface to the Reprinted Edition

When *The Newtonians and the English Revolution* first appeared fourteen years ago, some reviewers, in obvious discomfort, described the book as "bold." Others thought they spied not just the bold, but the positively dangerous. Some historians of science sought to dismiss the book for reasons of their own sociology, as not belonging to their academic discipline. Still others believed that the book "reduced" Newton's science to its social meaning. The ensuing controversy churned up that peculiar debate known among historians of science as the War between the Internalists and the Externalists. The vanguard of the latter were encamped in Edinburgh, and armed with *The Newtonians*, as well as with their own original work, they proclaimed a "strong program." For noncombatants that means a strong commitment to social and ideological elements as the central building blocks in scientific discourse. One of the most notable representatives of that school, Steven Shapin, wrote a lengthy critique of *The Newtonians* (*The Ferment of Knowledge,* Roy Porter and G.S. Rousseau, eds., Cambridge University Press, 1980). In that seminal essay Shapin drew out the implications of this book, along with others, for charting a new methodological course in Anglo-American historiography of science. As a result, *The Newtonians* entered the fray in what became the decisive battle in the Internalist/Externalist war.

By the late 1970s we were witnessing the maturation of social methods useful for the reading of historical texts about science. The social reading of science had come into its own. Partly inspired by that maturation, I attempted, with James R. Jacob, to draw out the

5

full implications of *The Newtonians* for the history of science and for
general history. We co-authored an essay entitled, "The Anglican
Origins of Modern Science: The Metaphysical Foundations of the
Whig Constitution" (ISIS, 1981).

The Newtonians describes the integration of Newton's science with
the interests of the Anglican church in the post-1689 English
constitutional order. It is a work of general history that seeks the
reintegration of the history of science into the narrative framework
of that general history. As such it possesses a double methodological
message: first, a socially anchored reading of scientific texts offers a
means to a larger end, that of arriving at historical knowledge about
the webs of discourse that permit scientific knowledge to be
generated. Secondly—and this is of particular importance for the
general historian—it offers new understanding of the power conferred by that discourse on the resulting knowledge generated about
what has been designated as nature. In so doing *The Newtonians*
implicitly argues that if these socially anchored methods are to be
successful, they must acknowledge the power of science, in short,
the ability of a scientific text, however socially constructed or
anchored, to exist in a dialectic with nature, with something other
than itself. Contrary to what some critics have misread into the
book, *The Newtonians* affirms the validity of mechanical science in
part because its natural philosophical discourse exists in time, in
relation to discourse that is, per force, socially and ideologically
present.

Most recently the social method of reading scientific discourse
has added new "technologies" to the collection of strategies for
decoding a text. The choice of terms is purposeful. We are seeing
the gradual entry of postmodern language theory into the historiography of science. With the assertion that all language masks and
reveals power, and that the dialectics revealed by a text are internal
to it, we have once again reopened, at least by implication, the issue
of the validity of modern science, of the direction it took at moments
in its historical development. We can put the matter another way: if
Newton can now be shown to have been a Newtonian what does this
say in response to the challenge of postmodern theory? The answer

implicit in *The Newtonians* should now be explicit: contextualization does not undermine the validity nor the "rationality" of science. Rather it refocuses the discussion onto the cultural meaning of that knowledge, the uses and abuses to which Western science has been put, toward questions about how and why the mechanical understanding succeeded, and the nature of the power derived from it. Into that enterprise all methods should be enlisted: gender theory, as well as language theory, ethnography, and not least the reconstruction and deconstruction of texts within a newly empowered historicism.

In the face of so much controversy any author would ask what should I have said differently. If I were to write the book again I would pedantically correct an error: the Richard Bentley writing to Evelyn about Milton (p. 152) is not the churchman but the bookseller of the same name. Not least, I would be even bolder and have more to say about Newton himself. In short, this is an unrepentent reprinting that does not apologize for having been bold, but does forgive the occasional impoliteness of some critics. It is the price the author paid for believing then and now that science possesses, then and now, an ideology integral to its being.

Acknowledgments

The manuscripts used in preparing this book came from many libraries, and I am especially indebted to the following and to their excellent staffs: the University Library, Cambridge, the libraries of King's College, Queens' College, and Trinity College, Cambridge; Christ Church Library, Oxford, where H. J. R. Wing was particularly helpful; and the Bodleian Library, Oxford. I am grateful to the Trustees of the British Museum for allowing me to use and to quote from their manuscript sources. Manuscripts at the British Library, the Lambeth Palace Library, the Leicestershire and Huntingdonshire Record Offices, the Victoria and Albert Museum, Dr. Williams's Library, the Royal Society of London, and the Bibliothèque Publique et Universitaire, Geneva, were used or quoted from by courtesy of these institutions. Also I am grateful to the late Olive Lloyd-Baker, C.B.E., for permission to use the Sharp MSS; the Trustees of the Will of the late J. H. C. Evelyn for permission to quote from the Evelyn Papers at Christ Church; and librarians Suzanne M. Eward, Norah Gurney, C. F. A. Marmoy, F.L.A., and Arthur E. Barker for their assistance.

I also had much encouragement from Virginia Harrison, librarian of the Newton Collection at Babson College, Massachusetts. The unpublished collections of some libraries have been consulted through correspondence, and I acknowledge the generous assistance of the Edinburgh University Library; the Borthwick Institute of Historical Research, York; the

Cathedral Library, Gloucester; and the Worcestershire Record Office. Other libraries provided essential printed sources: Olin Library, Cornell University; the Brotherton Library, University of Leeds, Yorkshire; and the New York Public Library.

My interest in the subject of this book dates back to my graduate years at Cornell University, where I was encouraged and assisted by Henry Guerlac, and also by H. G. Koenigsberger. My intellectual and personal debt to them is very great, and also to Dorothy Koenigsberger, who knows how laborious it can be to study what early modern Europeans called "esoteric wisdom."

At various stages the manuscript was read by other scholars; in every instance it has benefited from their comments. I am grateful to Henry Guerlac, I. B. Cohen, J. G. A. Pocock, L. Pearce Williams, Richard Schlatter, Gerald Straka, H. G. Koenigsberger, Ernst Wangermann, and Edward Pessen. I have also received helpful comments from Esmond de Beer, Esq., Frances Yates, Hillel Schwartz, David D. Brown, Wilfrid Lockwood, Henry Horwitz, and John Biddle.

In the final preparation of the text for publication I used the services primarily of Carol O'Connor, Judith Burns, and Julia Hilliard as typists, and John Saunders as research assistant. A portion of the research and much of the final preparation of the manuscript were made possible by generous research grants from the American Council of Learned Societies and the Research Foundation of the City University of New York. The Baruch College Scholar Assistance Program also made a contribution to this effort. Some of the material in this book first appeared in articles, and I thank various journals for their permission to adapt these articles or to reprint small portions from them: "John Toland and the Newtonian Ideology," *Journal of the Warburg and Courtauld Institutes,* 32 (1969), 307–331; (co-author Henry Guerlac), "Bentley, Newton and Providence (The Boyle Lectures Once More)," *Journal of the History of Ideas,* 30 (1969), 307–318; and in the same journal, "Millenarianism and Science in the Late Seventeenth Century" (1976), both by permission of the Journal of the History of Ideas, Inc.; "The Church and the Formulation of the Newtonian World-View," *Journal of European Studies*, 1 (1971), 128–148; (co-author W. Lockwood), "Political Millenarianism and Burnet's *Sacred Theory*," *Science Studies*, 2 (1972), 265–297; and finally "Early Newtonianism," *History of Science*, 12 (1974), 142–146. I am also grateful to Wilfrid Lockwood for his kind permission to use material in Chapter 3 on which he collaborated, and to Henry Guerlac for similar permission. This is but another example of his unfailing kindness toward me.

Another student of the seventeenth century has shared intimately in that process of personal and intellectual growth that makes the writing of a book possible and bearable. I am deeply grateful to my husband, Jim Jacob, for being committed to me and to the century that has fascinated us for so long. Other kindnesses from friends on both sides of the Atlantic brought relief from the hours spent in cold libraries. My friends in Yorkshire, New York, and, of course, in New Jersey should know that they are cherished and in part responsible for the completion of this book. It is dedicated to my mother, Margaret O'Reilly Candee, and for my father, Thomas W. Candee.

MARGARET C. JACOB

New York City

Abbreviations

B.L.	British Library, British Museum, London
Bodleian	Bodleian Library, Department of Western Manuscripts, Oxford
BTW	*The Theological Works of Isaac Barrow* . . . Edited for the syndics of the University Press by the Rev. Alexander Napier. Cambridge, 1859. 9 vols.
D.N.B.	*Dictionary of National Biography*
MSS ADD	Additional Manuscripts
P.R.O.	Public Record Office
TW	John Tillotson, *Works to which is prefixed the Life of the Author by Thomas Birch*. London, 1752. 3 vols.
U.L.C.	University Library, Department of Manuscripts, Cambridge

Note on Dates

Throughout this book, Old Style dates have been used, although I assumed the year to begin on January 1, not on March 25. However, where contemporaries used both Old Style and New Style in their dating of letters or sermons I have copied their practice.

The Newtonians and the
English Revolution
1689–1720

I received also much light in this search [of the Scripture] by the analogy between the world natural and the world politic. For the mystical language was founded in this analogy, and will be best understood by considering its original.

The whole world natural consisting of heaven and earth signifies the whole world politic consisting of thrones and people, or so much of it as is considered in the prophecy; and the things in that world signify the analogous things in this. For the Heavens with the things therein signify thrones and dignities and those that enjoy them, and the earth with all the things therein the inferior people, and the lowest parts of the earth, called Hades or Hell, the lowest and most miserable part of the people. Whence, ascending towards heaven and descending to the earth is put for rising and falling in honour and power. Rising out of the earth or waters, or falling into them, for the rising of any dominion or dignity out of the inferior state of people, or falling from the same into that inferior state. . . . Moving from one place to another [should stand] for translation from one office, dignity or dominion to another. Great earthquakes and the shaking of heaven and earth for the shaking of kingdoms so as to overthrow them. The creating a heaven and earth and their passing away, or, which is all one, the beginning and end of the world—for the rise and ruin of the body politic signified thereby.—Isaac Newton

Introduction

The problem considered here first presented itself to me
after I read certain standard works on the relationship between science and religion in seventeenth-century England.[1]
How could it be, these historians asked, that scientific ideas
were held and scientific methodology practiced by men such
as Robert Boyle and Isaac Newton, who were also deeply
religious, if occasionally unorthodox Christians? To their
surprise historians of ideas discovered that the question seemed
hardly to bother these pious proponents of the new science;
indeed, they seemed more concerned with the threat posed
by what they described as irreligion and atheism. They used
their science, often to the satisfaction of contemporaries, to
attack and refute the philosophies, both social and natural, of
their enemies. In so doing, these Christian and generally
Anglican proponents of the new science developed an explanation for the validity of Christianity, what they called
natural religion, or natural theology, which made their scientific ideas and natural philosophy respectable and also served

[1] For example, R. S. Westfall, *Science and Religion in Seventeenth
Century England* (New Haven, 1958); Charles Raven, *John Ray,
Naturalist: His Life Works* (Cambridge, 1950); G. N. Clarke, *Science and Social Welfare in the Age of Newton* (Oxford, 1938); Herbert Butterfield, *The Origins of Modern Science, 1300–1800* (London,
1949); G. R. Cragg, *From Puritanism to the Age of Reason* (Cambridge, 1950); Robert K. Merton, *Science, Technology and Society in
Seventeenth Century England* (New York, 1970, reprint); and the
earlier work of E. A. Burtt, *The Metaphysical Foundations of Modern Science* (New York, 1924).

as an indispensable prop for their version of Protestantism. The linkage established, in the late seventeenth century, between the operation of the physical order and the validity of Christianity was so indissoluble that only in the second half of the nineteenth century, when Darwin successfully challenged that explanation of nature, did the whole edifice of science-supported religion come crashing down. Darwin's science was lethal to religion only because in the second half of the seventeenth century a particular explanation of the natural order had gained acceptance as an indispensable support for liberal English Protestantism.

A careful reading of the literature of the seventeenth century revealed that the question of the compatibility of science and religion seemed no longer useful or even relevant. Instead, the crucial question became why did a particular natural philosophy, which I shall call the new mechanical philosophy, appear to certain influential Protestant thinkers as the only acceptable explanation of the natural order and hence the only valid support for what they understood Christianity to be?

The answer to that question unfolded not in the purely technical, scientific literature of the period, but in the sermons and treatises of certain late seventeenth-century churchmen. Some of these churchmen, such as John Wilkins and Isaac Barrow, were mathematicians and scientists of some repute; most of them were not. For the majority their interest in science meant reading the works of, or personally conversing with, the original scientific minds of their time, and possibly enjoying a fashionable membership in the Royal Society. What they knew about the new mechanical philosophy they culled, intelligently and purposefully, from its formulators, Robert Boyle and Isaac Newton, and to a lesser yet more complicated extent, from the Cambridge Platonists, Henry More and Ralph Cudworth. These churchmen then used the new mechanical philosophy in support of Christianity and in their assault on atheism and thus spread the ideas of the new science and its concomitant natural philosophy. Without the sermons of the first latitudinarians, science would have remained esoteric and possibly even feared by the educated but pious public.

The adjective "new" in the terms "new science" and "new mechanical philosophy" is used to distinguish the later science and mechanical philosophies from those of the first half of the seventeenth century. The new science championed by the latitudinarian churchmen constituted an explicit rejection of the mechanical philosophies of either Hobbes or Descartes, the science of the mid-century radicals, and the older Aristotelianism still extant in the universities and in some intellectual circles. The obvious question, of course, is why these liberal churchmen felt impelled to reject one science or natural philosophy and accept another. Historians of science have often presumed that the new mechanical philosophy triumphed in England simply because it offered the most plausible explanation of nature. It just may do that, but in my understanding of the historical process that made it acceptable the supposed correspondence of the new mechanical philosophy with the actual behavior of the natural order is not the primary reason for its early success.

This understanding of the triumph of the new science, in particular of the Newtonian natural philosophy, derives from a reading of the letters, sermons, and diaries of its first proponents, in particular of a handful of prominent Anglican churchmen. Out of those documents emerges a profoundly different explanation for the triumph of the Newtonian natural philosophy than that commonly offered by historians of science and of religion. These documents reveal that the Newtonian churchmen, like most seventeenth-century English proponents of one form of Christianity or another, were men with well-developed social and political interests. They possessed a vision of how the polity should operate to serve the church's interests, and that vision had been profoundly shaped by the mid-century Revolution. The politics and philosophies of the church's opponents in that struggle had forced these churchmen, most of whom came to intellectual maturity after 1660, to formulate a new version of Christianity and a new concept of how religion should serve the polity. We call them latitudinarians, a term to be discussed more fully in Chapter 1. The type of science most compatible for historical reasons with their social and religious philosophy was the natural philosophy first explicated by Boyle and Wilkins and in turn applied most successfully to the workings of the universe by Sir Isaac Newton. The latitudinarians accepted the new science and promoted it from their pulpits because it served their interests.

This social explanation for the triumph of Newtonianism in the late seventeenth century stresses what previous commentators have ignored—its usefulness to the intellectual leaders of the Anglican church as an underpinning for their vision of what they liked to call the "world politick." The ordered, providentially guided, mathematically regulated universe of Newton gave a model for a stable and prosperous polity, ruled by the self-interest of men. That was what Newton's universe meant to his friends and popularizers: it allowed them to imagine that nature was on their side; they could have laws of motion and keep God; spiritual forces could work in the universe; matter could be controlled and dominated by God and by men. Stability was possible without constant divine intervention; the spiritual order could be maintained; the church was necessary and essential; yet at the same time men could pursue their worldly interests. That, briefly stated, was what the world natural, explicated in the *Principia*, meant to churchmen who were primarily interested in promoting their vision of the "world politick."

This interpretation rests on the assumption, in some instances already documented by myself and other scholars,[2]

[2] See Chapter 1, opening remarks about Restoration science and natural philosophy.

that scientific ideas in seventeenth-century England could and did possess ideological significance. Because Hobbes and the radicals, among others, had used certain explanations of the natural order to support their political theories and politics, churchmen had to reject their natural philosophies along with their politics. In time Newton's vision of the universe would lend itself to interpretations different from those proposed by the latitudinarians. But in the late seventeenth century they seized upon his natural philosophy and the new science in general as a counterweight to what they called the atheism of the Hobbists, Epicureans, and radical freethinkers.

The linkage they forged between liberal Protestantism and early Newtonianism was never entirely broken during the eighteenth century. Witness the use to which it was put by that liberal proponent of godly capitalism Joseph Addison, or by that garrulous promoter of order and virtue Samuel Johnson.[3] William Blake was not being simply rabid when he assaulted Newton as the symbol of a market society, based on technology and empire, which oppressed him.[4] The latitudinarian proponents of early Newtonianism had succeeded in resting their social ideology on the model provided by the Newtonian universe.

As perhaps in all success there is historical irony. The society that the latitudinarians wished to create was to be Christian and godly in the biblical sense of those terms. Their vision of history had been conditioned by the Reformation, and they believed themselves to be preparing Englishmen for the millennial paradise. One of the more curious topics to be discussed in this book is the millenarianism of these churchmen. Lest we imagine them to be Enlightenment men—as so

[3] See the admirable study by Edward A. Bloom and Lillian D. Bloom, *Joseph Addison's Sociable Animal* (Providence, R.I., 1971); and Richard B. Schwartz, *Samuel Johnson and the New Science* (Madison, Wis., 1971).
[4] The subject is dealt with in Donald Ault, *Visionary Physics: Blake's Response to Newton* (Chicago, 1974), but the political and ideological aspect of Blake's response is not treated.

many historians perilously have done with Newton—we must keep in mind that their underlying assumptions about the meaning of history, and therefore of politics, were profoundly Christian. The latitudinarians are transitional figures; they flourished in that period from 1680 to 1720 when the natural philosophies of the seventeenth century were being applied to a society increasingly dominated by market values—in short by capitalism. They never, however, accepted or even fully understood the practices, life styles, and values of the wealthy London parishioners to whom they mostly spoke.

In the chapters that follow I will try to explain who both the first and second generations of latitudinarians were and to describe the natural philosophies they rejected and why they rejected them. Most of my attention is focused on the second generation of latitudinarians, the so-called Newtonians. In Chapter 2, I describe how the Revolution of 1688–1689 affected them and compelled them to chart the intellectual course that led to their acceptance of Newton's natural philosophy. An explication, based on printed sources, of the relationship between the Newtonians' vision of society and their acceptance and use of Newton's natural philosophy occurs in Chapter 5. Finally, I will explore the beliefs and activities of their opponents, the freethinkers and enthusiasts. We do not know exactly how many of them there were, but certainly most seventeenth-century churchmen regarded them as a serious threat and that is sufficient reason for dealing with them at some length.

I have employed an interdisciplinary method purposefully to merge subjects traditionally reserved for the history of science, or church history, or intellectual history. For if science in the seventeenth century possessed social relations, and if churchmen had political interests, and if scientists could also be millenarians, then our approach to these problems must be interdisciplinary. If the Newtonian natural philosophy gained acceptance and popularity because it effectively supported a particular social ideology, then this methodology is the only route open to the historian who wishes to uncover that historical relationship.

From this interdisciplinary perspective scientific thought and natural philosophy in the late seventeenth century receive historical treatment in their social context, in relation to the political interests of their Anglican explicators and proponents. Never is it implied, however, that social and political conditions account for the technical problems that were posed or solved by these natural philosophers. All intellectual activity is not about, or determined solely by, social relations and problems. Yet, if I seek to establish anything, it is the vital role played by social and political issues in the formulation and acceptance of the new science, in particular of the Newtonian natural philosophy. Of course, the proponents of the new science were churchmen, both lay and clerical, whose sensibility was profoundly Christian. There is a point in matters of religious sensibility beyond which it is hazardous for the historian of ideas to comment. The latitudinarians were public men, and I have tried to understand their public roles in relation to their private beliefs. The psychology of their personal piety and the nature of their intimate religious emotion are assumed to be matters beyond the scope of this essay. Both science and religion can be, at times, just that—the search for a technical understanding of natural phenomena, or a personal system of ethical standards coupled with the sense of a supernatural power. At other times and even simultaneously, both can be perceived by churchmen, scientists, and therefore by historians as the foundations of a social ideology. This book attempts to explore one such scientific and religious ideology, the Newtonian, in its social context, that is, in relation to that matrix of social and political change best described as the English Revolution.

CHAPTER 1

Latitudinarian Social Theory
and the New Philosophy

The social meaning given to Newton's science by his
friends and followers in the church had antecedents in the intellectual program of certain Restoration divines. This first
generation of latitudinarians established the precedent of link-
ing the new mechanical philosophy with specific social and
political interests. That precedent may be dealt with in a general way, acknowledging, but not dwelling on, the complexities of Restoration natural philosophy and religion. Among
Protestant and Anglican proponents of the new science, such
as Robert Boyle and Henry More,[1] substantial philosophical
differences existed. Yet all the latitudinarians, Boyle, More,
John Wilkins, John Tillotson, Isaac Barrow, and Simon Patrick, gave the same social meaning to natural philosophy. All
used the new mechanical philosophy, that is, their vision of
the natural world, to support a political world where private
interest would enhance the stability of the public weal and
Anglican hegemony would rest secure.

What is said in this chapter about Restoration science and
society is intended primarily to set the historical context for
the ensuing discussion of Newtonianism. From that perspec-

[1] See Robert Boyle, *A Defense of the Doctrine Touching the
Spring and Weight of the Air* (London, 1662); *A Continuation of
New Experiments Physico-Mechanical, Touching the Spring and
Weight of the Air* (Oxford, 1669); *An Hydrostatical Discourse Oc-
casion'd by Some Objections of Dr. Henry More* (London, 1672);
Marjorie H. Nicolson, ed., *Conway Letters* (New Haven, 1930), 269.

tive the most significant intellectual achievement of Restoration natural philosophers—men like Wilkins, Boyle, and Barrow—was the articulation of a mechanical philosophy that required God's active participation in the workings of nature. In this Christianized philosophy of science the universe was composed of matter in motion; matter possessed an atomic structure, and the atoms collided in empty space. Motion, however, was separate from matter, imposed on atoms as well as on the objects they formed by outside, spiritual forces. The design and harmony so evident to Restoration natural philosophers existed only because the providential deity supervised every operation of nature. He did so through rules that expressed his will, and the task of science was to discover and explicate the rules that govern nature, in effect to explain to man the operations of providence in creation.

Restoration philosophers promoted their mechanical philosophy because they believed it to be an essential underpinning for true Christianity. Their understanding of the importance of Christianity and therefore of the church had been shaped by the mid-century Revolution and by their opposition to the philosophies that had gained expression and acceptance during the late 1640s and 1650s. The proponents of the new mechanical philosophy explicitly pitted it against rival natural philosophies which they believed to be pernicious and dangerous to the social and political order restored in 1660. In short, the new mechanical philosophy from its very inception possessed social and political significance.

That significance can be understood only if we see the social implications inherent in the rival philosophies that Boyle, Wilkins, and Cudworth sought to undermine. In the period from 1640 to 1660, when censorship broke down and the polity was shaken by revolution, every conceivable theory about religion, society, nature and God could be read by the literate population or preached to the nonliterate. During the English Revolution particular natural philosophies came to be associated with particular political and social theories, or occasionally with political behavior.[2] For instance, radicals such as the Levellers preached and wrote about nature in a way that can best be described as pantheistic and materialistic. Alternatively, the most important political philosopher of the period, indeed of the century, Thomas Hobbes, wrote about man and society in the same way that he wrote about nature—Hobbes's materialism is all of a piece.

Throughout the seventeenth century, thinkers of whatever philosophical or political persuasion assumed that some sort of relationship, of varying degrees of causality or simply of intimacy, existed between the world natural and the moral and social relations prevailing or desired in the "world politick." Operating, therefore, with basically the same set of assumptions as those of their disparate protagonists about the relevance of the natural order to the political order, churchmen assumed that to accept the radicals' vision or Hobbes's vision of the political order meant to accept their vision of the natural order, or vice versa. In some instances, especially in the case of the radicals, the linkage between pantheistic materialism and theories about social change and political reorganization was breakable. But that provided little reassurance for those who did not want the radicals and the enthusiasts to gain control, to dissolve property rights, and to destroy all forms of church organization. Opposition to the radicals inevitably meant opposition to everything they stood for, in philosophy as well as in politics. For churchmen who also feared—for different reasons—Hobbes's analysis of man and society, his materialism was dangerous and pernicious, and they had to find an alternative to it. Robert Boyle explicitly stated that it had been suggested to him that one way of refuting Hobbes's "dangerous opinions about some important,

---

[2] See J. R. Jacob, *Robert Boyle and the English Revolution* (New York, 1977); P. Rattansi, "Paracelsus and the Puritan Revolution," *Ambix*, 11 (1964), 24–32; and C. B. Macpherson, *The Political Theory of Possessive Individualism, Hobbes to Locke* (Oxford, 1962).

if not fundamental, articles of religion" would be "to shew, that in the Physics themselves, his opinions, and even his ratiocinations, have no such advantage over those of some orthodox Christian Naturalists"; and that was precisely what Boyle attempted to do. John Wallis, a founder of the Royal Society and a latitudinarian in matters of religion, also believed that to refute Hobbes's science would aid in refuting his opinions in philosophy and theology.[3]

Although in seventeenth-century England certain theories about the world natural seemed most compatible with certain other theories about the "world politick," that linkage could be, and occasionally was, broken. I do not mean to imply that only certain natural philosophies could ever support certain political and social theories, that, for instance, Newtonianism could only support the social vision to which churchmen gave it application. My argument is simply that in the seventeenth century certain natural philosophies became linked with certain political ideologies for historical reasons. If circumstances changed, as they inevitably must, then models of the natural order might be used in altogether different ways (or possibly not used at all) as part of an individual's or a group's ideology. For the latitudinarians, as we shall see, the new mechanical philosophy served as the foundation for a social ideology with a dual purpose: to secure and legitimize church and state against the threats posed by radicals, enthusiasts, and atheists and also to reform this established order. These churchmen and natural philosophers wanted to Christianize social relations in a market society that they believed already existed in England. They wanted also to provide for religious peace in such a society by instilling their version of liberal Protestantism, and finally, they sought to harmonize human history with

[3] Robert Boyle, Preface, *An Examen of Mr. T. Hobbes His Dialogus Physicus de Natura Aeris* (1662) in *Works*, I (London, 1744), 119; Peter Toon, ed., *The Correspondence of John Owen (1616–1683)* (Cambridge, 1970), 87–88, Wallis to Owen, Oct. 10, 1655. I owe these references to J. R. Jacob.

the providential plan and thereby to prepare the way for the coming millennium. The political and natural philosophies of Hobbes and of the radicals threatened the very survival of the church, but more immediately and realistically their acceptance jeopardized all the hopes the latitudinarians nurtured for reform.

The debate about whether or not Hobbes derived his mechanical explanations of human behavior from the new science or from his experiences cannot be settled here. In conjunction with other commentators on Hobbes, including his own contemporaries, I want to assert the fundamental union of his science-based materialism and his political theory. Because he analyzed men as mechanisms in motion, as essentially self-moved matter, he could postulate that morality sprang from their motions and not from any source outside themselves. Because spiritual principles that ultimately derive from the deity played no part in Hobbes's materialism, the traditional explanations of political obligation—that it was based on reason or obedience to God's will—played no part in his political theory.[4] He would have man's competitive nature, derived from his relations in a market society, achieve recognition and acceptance within the context established by an absolute sovereign capable of restraining and channeling human aggression and competition. Although Hobbes believed in divine revelation and therefore in providence,[5] he separated the providential order from the "world politick" and the world natural and presumed that the dictates of providence were inevitable, matters to be accepted by faith, yet equally irrelevant to the social order. In *Leviathan* (1651) as well as in his other writings, Hobbes addressed himself to the market society and therefore to self-interest; indeed he was concerned about the very social and moral problems that vexed the

[4] Macpherson, 78–79.
[5] J. G. A. Pocock, "Time, History and Eschatology in the Thought of Thomas Hobbes," in J. H. Elliott and H. G. Koenigsberger, eds., *The Diversity of History* (Ithaca, 1970), 149–198.

latitudinarians. Perhaps for that reason they feared him as they feared no other philosopher.

After the Restoration the immediate political threat posed by the radicals seemed to lessen, yet political and philosophical radicalism survived. It surfaced during the Exclusion crisis and Monmouth's rebellion; and the leaders of the 1688–1689 Revolution feared its resurgence and did their work quickly in part to prevent any subversion from the left of their revolution. In the heated political life of England after 1688–1689, the heirs of the Commonwealth tradition, meek fellows compared to the many radical groups of the 1650s, once again launched their attack on the established order in church and state. When early eighteenth-century churchmen waged war with the freethinkers, they imagined that they saw the specter of mid-century radicalism.

Indeed there are connections between the philosophical and ideological positions of someone like John Toland (1670–1722) and the radicalism of mid-century. Both generations articulated a natural philosophy that could be described as pantheistic materialism. In a later chapter the thought and activities of the early eighteenth-century radicals will be discussed in detail. For the present, it should be stressed that the battle lines between latitudinarians and radicals were drawn up in the 1650s. During the English Revolution, groups like the Ranters or philosophers like Gerrard Winstanley equated God with nature or with reason.[6] For Winstanley, "the whole creation . . . is the clothing of God," and from that principle the radicals could argue for the equality of all things in creation, the futility of priestcraft and property, and the necessity for science based on Paracelsian and magical principles. Science would serve the people by using the people's science—astrology, alchemy, natural magic[7]—which would be studied systematically and its useful principles applied to un-

[6] Christopher Hill, *The World Turned Upside Down* (London, 1972), 112–114.
[7] Ibid., 233–236.

raveling the secrets of creation and to the practical arts, medicine, pharmacology, and architecture. The rejection of the magical tradition and its subsequent transformation by philosophers like Boyle and Newton into what we call modern science occurred in a political context that repudiated and feared the politics and therefore the science of the radicals. That is not to say that the magical tradition did not play a fundamental role in the origins of modern science. We know that the Hermetic tradition could and did serve many masters. When churchmen championed the new mechanical philosophy, however, they did so in part as a viable alternative to the pantheism of the radicals.

In the 1650s and early 1660s, certain church intellectuals such as the Cambridge Platonists Henry More and Ralph Cudworth and clerics such as Simon Patrick and Isaac Barrow thought that Descartes's philosophy, if combined with a modified Platonism, could provide a guarantee against materialism. Gradually, however, disenchantment set in as they came to fear that Descartes's "mechanical pretensions" threatened religion.[8] The reception and ultimate rejection of Cartesianism in England are complex problems,[9] yet one theme remained constant in that process. At first the supporters of Anglican Protestantism saw Descartes as an ally, and their subsequent rejection of his natural philosophy rested not simply on the inconsistencies and inaccuracies of his scientific explanations; it also had much to do with their fear that his mechanical philosophy led straight to materialism. The rejection of Descartes in England was completed by Newton, who consciously repudiated his science and his natural philosophy.

Increasingly during the Restoration, English natural philosophers adopted a modified version of the mechanical philoso-

[8] Charles Webster, "Henry More and Descartes: Some New Sources," *British Journal for the History of Science*, 4 (1969), 359–364.

[9] See Robert Kargon, *Atomism in England from Harriot to Newton* (Oxford, 1966), and other secondary sources cited therein.

phy. The originators of this philosophy, Henry More, Ralph
Cudworth, Robert Boyle, John Wilkins—to name only the
most prominent—wrote voluminously. They engaged in public
controversy with Hobbism and sectarianism, and their ideas
were in turn used by other churchmen, often less skilled or
less original in natural philosophy and mathematics. The
preaching of natural philosophy became fashionable during
the Restoration, and beginning in 1692 with the Boyle lectures
of Richard Bentley, the practice achieved historical fame. In
his lectures, the Newtonian natural philosophy received its
first and highly simplified explanation aimed at a nonspecialist
audience, most if not all of whom, like the lecturer, could
never have followed the mathematical reasoning inherent in
Newton's *Principia* (1687).

Bentley's rather minimal mathematical and scientific knowledge was typical of the church circle he represented. As
clerics they were involved in the day-to-day activities of their
offices, as well as in ecclesiastical and state politics. If their
preaching and politics were to gain recognition, and thereby
patronage and preferment, they had to preach about what
interested their congregations and patrons. Bentley and the
latitudinarians were distinguished from other churchmen by
their interest in the new science. At first they were unique
within the church, and in the early Restoration they stood out
as a dissident minority. As their views became accepted, the
latitudinarians advanced to high ecclesiastical offices, and by
the 1690s they controlled most of them.

The first generation of latitudinarians, or low-churchmen
as they were called, gave intellectual respectability and notoriety to the new mechanical philosophy, and they fashioned
an explanation of its relevance to the social and political order
which was adopted and expanded upon by the Newtonians.
No single definition can or should be given for latitudinarianism, and its adherents often differed among themselves on
some topics, in particular party politics. Any characterization
of them as Whigs is inaccurate. By and large, and particularly

after the Revolution of 1688–1689, the latitudinarian bishops tended to side with the court Whigs,[10] but here too Thomas Sprat, bishop of Rochester and author of the *History of the Royal Society*, and John Sharp, archbishop of York, were latitudinarian and Tory.[11] Latitudinarianism was primarily a religious and ecclesiastical position and can best be seen as yet another form of English Protestantism. Indebted, of course, to the main sources of Protestantism, to Luther, indirectly to Calvin, and, of course, to Hooker and Chillingworth, the Anglican moderates of the Restoration nonetheless fashioned their own Protestantism, or natural religion as they liked to call it.

Of the clerical and intellectual leaders of this first generation, undoubtedly John Tillotson (1630–1694) ranks foremost. The son of a clothier, he began his clerical life at Cambridge in the 1650s as a Presbyterian who finally conformed only at the Restoration.[12] After the Revolution of 1688–1689 he was William's personal choice for the archbishopric of Canterbury at the recommendation of Gilbert Burnet, and in that capacity he made many enemies among high-churchmen. Typical of his church circle, Tillotson was a Fellow of the Royal Society. having been proposed by Seth Ward, bishop of Salisbury. Tillotson was also on good terms with Edmond Halley in the early 1680s,[13] and, of course, it was Halley who urged Newton

[10] F. G. James, "The Bishops in Politics, 1688–1714," in William A. Aiken and Basil D. Henning, eds., *Conflict in Stuart England* (London, 1960).

[11] For Sprat see J. Hough, *Table Talk and Papers*, in *Collectania*, 2d ser. (Oxford, 1890), 389; and G. V. Bennett, "Conflict in the Church," in Geoffrey Holmes, ed., *Britain after the Glorious Revolution, 1689–1714* (London, 1969), 165.

[12] For his patrons see MSS ADD 9828, ff. 125–126; MSS ADD 17017, ff. 142, 145–146, B.L.; J. Tillotson, *The Works*, ed. T. Birch (London, 1752), I, xi, cxv, hereafter cited as *TW*.

[13] MSS ADD 17017, ff. 145–146; Sloane 179, a. XVII cent., f. 153; MSS ADD 17017, f. 143, B.L.; L. M. Hawkins, *Allegiance in Church and State: The Problem of the Nonjurors in the English Revolution*

to publish the *Principia*. Tillotson's social ties extended out of the church and university to include London businessmen, his brother Joshua, and Thomas Firmin, a progressive entrepreneur interested in putting the poor to work.[14]

Tillotson associated with almost every prominent latitudinarian. He married John Wilkins' daughter, and her father's thought and career offer a case study of the relationship between latitudinarianism and science.[15] Also among Tillotson's friends was Isaac Barrow, the most brilliant mathematician of the generation before Newton and his teacher. Wilkins encouraged Barrow's career as did Tillotson, who collaborated with him in the writing of at least one important sermon and edited his posthumous tracts and sermons.[16] Into this circle came Simon Patrick, who was on good terms with both Tillotson and Barrow. Patrick attempted one of the first public defenses of the latitudinarian position, and after being passed over for many years he eventually obtained a bishopric in 1689.[17] Also elevated at that time was John Moore, whose exercise of patronage or general encouragement aided the careers first of Isaac Barrow, then of the prominent New-

(London, 1928), 131-159; MSS Cole 5822, f. 64, B.L.; MSS ADD 4236, ff. 230, 233, 227, B.L.; and *TW*, 65 et seq.

[14] A. Trevor, *The Life and Times of William, the Third* (London, 1835-1836), II, Appendix I, 472; H. W. Stephenson, "Thomas Firmin, 1632-97," 3 vols. (Ph.D. diss., Oxford University, n.d.).

[15] Barbara Shapiro, *John Wilkins, 1614-1672: An Intellectual Biography* (Berkeley, 1969); J. R. Jacob and M. C. Jacob, "Scientists and Society: The Saints Preserved," *Journal of European Studies*, 1 (1971), 87-92.

[16] P. H. Osmond, *Isaac Barrow: His Life and Time* (London, 1944), 96-97; I. Barrow, *Theological Works*, ed. A. Napier (London, 1859), I, lxxi; Irene Simon, "Tillotson's Barrow," *English Studies*, 45 (1964), 193-211, 273-288. Barrow's works hereafter cited *BTW*.

[17] Seth Ward to Patrick, regards sent by Barrow, June 17, 1671, MS 73, no. 25, Patrick's Letters, Queens' College, Cambridge; for Tillotson and Patrick, MSS ADD 20, f. 7, U.L.C. I have accepted Patrick as the author of S. P., *A Brief Account of the New Sect of Latitude-Men* (London, 1662).

tonians William Whiston and Samuel Clarke.[18] Joseph Glanvill also belonged to this circle and during the Restoration became one of the chief defenders of the new science. There were many lesser lights who are nonetheless important for our understanding of certain aspects of latitudinarianism. Edward Fowler, bishop of Gloucester (1691–1714), served as propagandist and polemicist; William Lloyd, bishop of St. Asaph and finally of Worcester (1700–1717), interpreted the Scriptural prophecies as best he could; Hezekiah Burton negotiated with the Dissenters; and Thomas Burnet, for a time master of Charterhouse, attempted to synchronize natural history with ecclesiastical and political history, past and future.[19]

One latitudinarian who earned almost universal admiration as a lay churchman and scientist was Robert Boyle, and his thought affords perhaps the most complex example of the relationship between natural religion and natural philosophy.[20] The latitudinarian debt to Boyle was so profound that it would be tedious to document. Perhaps the first example of the use of Boyle's ideas by a churchman appears in the first edition of Edward Stillingfleet's *Origines Sacrae* (1662), where Boyle's natural philosophy is adopted to answer political and religious questions.[21] By 1697, Stillingfleet had come to accept Newtonian explanations of phenomena wholeheartedly and revised the *Origines* accordingly.[22] Boyle's influence among the latitudinarians was institutionalized after his death in

[18] Osmond, 132; William Whiston, *Historical Memoirs of the Life of Dr. Samuel Clarke* (London, 1730), 7–8.

[19] Jackson I. Cope, *Joseph Glanvill, Anglican Apologist* (St. Louis, 1956); M. C. Jacob and W. A. Lockwood, "Political Millenarianism and Burnet's *Sacred Theory*," *Science Studies*, 2 (1972), 265–279.

[20] See J. R. Jacob, *Boyle*; cf. his "Robert Boyle and Subversive Religion in the Early Restoration," *Albion*, 6 (1974), 275–293.

[21] Stillingfleet, *Origines Sacrae* (London, 1662), 424–428, 458. I owe this reference to J. R. Jacob.

[22] Richard Popkin, "The Philosophy of Bishop Stillingfleet," *Journal of the History of Philosophy*, 9 (1971), 303–319; and E. Stillingfleet, *Works* (London, 1709), II, 116.

1691 in the Boyle lectures, monthly sermons given at Boyle's request for which his will provided an annual stipend. These lectures were controlled by the latitudinarians and in particular by Thomas Tenison, archbishop of Canterbury (1695–1714), and John Evelyn, close friend of Boyle. In the early eighteenth century the lectureship served as a podium for latitudinarian thought and for Newtonianism.

The main trustees of the lectureship, Tenison and Evelyn, had scientific interests but few, if any, scientific achievements; Evelyn was a founder of the Royal Society and Tenison showed an interest in the new philosophy as a young man.[23] Like almost all the other latitudinarians, he too was a member of the Royal Society. Since Evelyn's diary serves as a veritable weather vane of latitudinarian sentiment on political matters, his attitudes will receive more attention in this study than they are commonly accorded. He, Stillingfleet, and, as I shall argue, quite possibly Newton, sponsored the career of Richard Bentley, the first Boyle lecturer.

When Sir Isaac Newton associated with churchmen, they were the latitudinarians. Indeed, his religious and ecclesiastical thinking so generally reflects latitudinarian positions that it is not unique. Tenison supposedly asked him to take orders, but he refused, claiming that he could do more for the church as a layman. During the Revolution of 1688–1689, Newton conferred with both churchmen and secular politicians.[24] In the reign of Anne he frequented low-church gatherings at the home of William Wake.[25] Although this study does not deal directly with Newton himself, except where he is representative of

[23] H. Oldenburg, *The Correspondence*, trans. and ed. A. R. Hall and M. B. Hall (Madison, Wis., 1971), VIII, 344–348.

[24] L. T. More, *Isaac Newton: A Biography* (New York, 1934), 608; *TW*, I, xlv. On January 17, 1688/9, William of Orange dined with Newton, Sawyer, and Mr. Finch according to Roger Morrice, "Entering Books," II, f. 429, Dr. Williams's Library, London.

[25] "Wake's Diary," Nov. 8 and 13, 1706, MS 1770, Lambeth Palace Library. Newton later quarreled with Wake, Newton MS 130, King's College Library, Cambridge.

commonly held latitudinarian views or aided the latitudinarians, it should help to create a context within which his religious sensibility and natural philosophy can best be understood.

With very few exceptions the latitudinarians were Cambridge men, and they may have drawn many of their notions in natural philosophy from the Cambridge Platonists, especially from Ralph Cudworth and Henry More, whom in almost every case they knew personally. Simon Patrick was also close to the Platonist John Smith; Thomas Burnet studied with Cudworth; and John Tillotson was particularly devoted to Benjamin Whichcote. Much has been made of Newton's intellectual debt to the Cambridge Platonists.[26] My intention in this chapter is not to discuss intellectual influences on the first generation of latitudinarians. Rather it is necessary to establish historical background for the social context of Newtonianism and therefore to discuss the social and intellectual world of the Restoration latitudinarians, especially of those who became most closely associated with the Newtonians. Barrow taught the young Newton,[27] Wilkins is too influential not to be discussed, Moore was patron and friend to Bentley and Clarke, and Tillotson as archbishop after the Revolution dominated church policies.

These Restoration churchmen shared certain basic convictions which were the tenets of their natural religion: rational argumentation and not faith is the final arbiter of Christian belief and dogma; scientific knowledge and natural philosophy are the most reliable means of explaining creation; and politi-

[26] D. D. Brown, "An Edition of Selected Sermons of John Tillotson (1630–1694) from MS Rawlinson E. 125 in the Bodleian Library" (M.A. thesis, University of London, 1956), lii; J. E. McGuire and P. Rattansi, "Newton and the Pipes of Pan," *Notes and Records of the Royal Society of London*, 21 (1966), 108–143. First discussed in general terms by E. A. Burtt (see Introduction, n. 1), and also by A. Koyré, *From the Closed World to the Infinite Universe* (Baltimore, 1957).

[27] E. W. Strong, "Barrow and Newton," *Journal of the History of Philosophy*, 8 (1970), 155–172.

cal and ecclesiastical moderation are the only realistic means
by which the Reformation will be accomplished.

Latitudinarian moderation dramatically contrasted with the
circumstances within which it was conceived. Civil war,
sectarianism, political radicalism, and Catholic dominance on
the Continent confronted the churchmen who forged latitudinarianism. The compromising and moderate nature of
their response to social and political upheaval has led some
commentators naively to assume that the latitudinarians could
only have been otherworldly men who retreated from their
hostile environment. These Christian humanists, it has been
argued, "felt pent in alike by narrowness of ritual and by
narrowness of creed, and they cried out for room and air. To
these expansive souls the atmosphere both of triumphant
Puritanism and of triumphant Anglicanism was stifling." Once
alienated, an important latitudinarian like Henry More retired
"within . . . [and] lived in a great calm. His environment
mattered little."[28] From the vantage point of this supposed
noninvolvement More embraced the new philosophy of
Descartes—only later to reject it—wrote cabalistic poetry and
denounced atheism and enthusiasm. Although one might plausi-
bly argue that More's temperament was quietistic and retiring,
that interpretation hardly holds for More's followers and
friends. Contrary to much that has been said by historians
about these moderate churchmen, they were intimately involved in contemporary political and ecclesiastical life. Their
use of science, reason, and moderation was intended to maintain and increase the church's domination over the religious
life of the nation.

If reason, science, and moderation worked their cure, the
latitudinarians foresaw society governed by reason and divine
law as interpreted by the church. Acquisitiveness and self-
interest—the motivations churchmen believed paramount in
their society—would not cease; rather, self-interest would be

[28] Edward A. George, *Seventeenth Century Men of Latitude* (London, 1909), 6, 109.

made enlightened and private interest would flourish in the service of public interest. Natural religion would curb rapacious greed and render its practitioners into godly men. As natural religion came to prevail, so too would the church. The ascendancy of a unified Protestant church in England, made possible by a broad and reasonable version of Christianity, would herald the renewal of ecumenism in Europe and usher in a reunification of European Protestantism. All these events had been foretold in the Scriptural prophecies, and they were part of the divine providential plan. That plan moved inexorably toward fulfillment, and the latitudinarians saw their natural religion as an ideology in the service of providence.

Science functioned as an essential ingredient of the latitudinarian social ideology. From the pulpit, the latitudinarians translated what they had read in the natural philosophers to a wide and captive audience. This is not to say, however, that the translation of natural philosophy into simple concepts easily grasped by an educated audience engaged in commerce or university life or city affairs met with indifference or disapproval from scientists or natural philosophers. On the contrary, the founders of the Royal Society, according to Sprat's history of that institution, heartily approved and shared the goals of the latitudinarian churchmen.

Confirmation of the Society's approval lies in that masterful piece of propaganda, Thomas Sprat's *History of the Royal Society* (1667). The Society commissioned Sprat's account, Wilkins encouraged his project, and Boyle published views similar to Sprat's.[29] Ostensibly Sprat wished to put forward the image of science as a retreat from worldly considerations, specifically religious and political. A cursory reading of his

[29] Shapiro, 203–204. See Part 1 of "Some Considerations Touching the Usefulness of Experimental Natural Philosophy" in Thomas Birch, ed., *Works of the Honourable Robert Boyle* (London, 1744); I owe this point to J. R. Jacob; see his "Restoration, Reformation and the Origins of the Royal Society," *History of Science*, 13 (1975), 155–176; and Barbara Shapiro, "Science, Politics and Religion," *Past and Present*, no. 66 (1975), 137.

History may encourage the belief that scientific pursuits did represent such a retreat. Science seemed to afford a new and independent criterion of truth, divorced from social pressures and devoid of ideological content. Read carefully, however, Sprat's account of the Royal Society—its origins and aims— reveals that its supporters grasped the meaning of their scientific activity largely by reference to political and religious goals. Sprat tells us that natural philosophy "is a religion, which is confirmed, by the unanimous agreement of all sorts of Worships."[30] The study of nature transcends sectarian and ecclesiastical disputes and produces a profound belief in the power, wisdom, and goodness of the creator and an immediate sense of the order of creation. Science instills a private faith which in respect to Christianity resembles the porch to Solomon's Temple—and the direct beneficiary of its support can only be the English church. Science repudiates religious enthusiasm: "For such spiritual Frenzies, which did then bear Rule [during the Interregnum] can never stand long, before a clear, and a deep skill in Nature."[31] Sprat uses the scientific activities of the Royal Society and its empirical methodology as a weapon against the "inner light" or Anabaptist view of religion and church structure that dominated the thinking of the large numbers of radical sects that had threatened not only the church but also the social and political hierarchy.

Because Sprat's understanding of the religious meaning of science was conditioned by the Revolution, it was profoundly political. Sprat offers the moderation instilled by the pursuit of true experimental science as a force for political moderation. The task of science is so vast that "it cannot be perform'd without the assistance of the Prince":

It will not therefore undermine his Authority whose aid it implores: that [science] prescribes a better way to bestow our time, than in contending about little differences, in which both the

[30] Thomas Sprat, *History of the Royal Society*, ed. Jackson I. Cope and Harold W. Jones (St. Louis, 1958), 82.
[31] Ibid., 63–64, 82, 54.

Conquerors and the Conquer'd have always reason to repent of
their success: [science] shews us the difficulty of ord'ring the
very motions of senseless and irrational things; and therefore how
much harder it is to rule the restless minds of men: [science]
teaches men humility, and acquaints them with their own errors;
and so removes all overweening haughtiness of mind, swelling imaginations, that they are better able to manage Kingdoms than
those who possess them. This without question is the chief root
of all the uneasiness of Subjects to their Princes. The World
would be better govern'd, if so many did not presume that they
are fit to sustain the cares of Government.[32]

The political function of science, as rendered by Sprat, becomes the maintenance of order and stability. Scientific activities practiced by every level of society, by mechanics, and especially by the rich and great, will reap unimagined rewards.
For science serves trade and industry, and together they ensure the growth and maintenance of empire within the British
Isles and overseas.[33] Sprat's brilliance lay in his ability to use
science to help legitimize his view of the restored social order
and simultaneously to put his social and religious ideology into
service as a guarantor for the experimental approach to science
championed by the Royal Society.

If science can serve to promote stability, and if through
natural religion men settle their doctrinal disputes, then the
empire Sprat envisaged would be truly Christian and Protestant. Both the church and the Royal Society implement the
Reformation, one in the area of religion, the other in philosophy, and "the Harmony between their Interests andTempers"
should ensure a mutual cooperation that will unite all Christians.[34] The peace and joy conferred by the triumph of true
religion and true science will be matched only by the power
and prosperity extracted from industry, trade, and empire
through the applications of science. Sprat's message was re-

[32] Ibid., 429–430.
[33] Ibid., 403, 86–88, 131.
[34] Ibid., 370–371, 427.

peated by churchmen: science encourages political moderation while at the same time providing a new understanding of creation and thereby providing a foundation for natural religion. Because of its reliance on true methodology, on experimentation, and on reason, the science of the Royal Society engenders a natural religion that necessarily conforms to true Christianity as revealed by God in Scripture. Churchmen naturally found true Christianity to be both Anglican and Protestant.

Some churchmen doubted, however, that latitudinarian natural religion conformed to true Christianity. They saw in latitudinarianism a profound threat to the church's doctrinal purity and political hegemony. The first generation of low-churchmen met with considerable opposition. Indeed this opposition distinguishes them from the Newtonians, not in ideology, but in rhetoric and polemics. The first generation of latitudinarians felt compelled to provide arguments to prove the importance of the new experimental science as the key to understanding the natural world. By the 1690s, however, the latitudinarians were in political ascendancy; science, thanks to Newton, Boyle, and others, had become more respectable; and the Newtonians could concentrate their energies on refining and developing the social ideology initiated by Barrow, Wilkins, Boyle, and others and on applying it to the market society with renewed vigor.

One of the main reasons why the first generation of latitudinarians experienced hostility from within the church derived from their continued residence in England during the Interregnum.[35] Some had been students, others became clerics and accepted Presbyterian ordination with varying scruples. At the Restoration, the Laudian faction, many of whom had gone into exile, returned to power. Unlike the latitudinarians, the Laudian exiles had refused any compromise with a Commonwealth that abrogated the church's legal dominance. In

[35] Barrow did travel on the Continent for a time. He was the most avowedly Royalist of the group.

1662 they fashioned a religious settlement that relegated even the most moderate Dissenters to the status of an illegal minority.

Publicly the restored church was united, its Laudian and moderate wings engaged in the task of reconstituting the church's administrative structure. Indeed that machinery was reconstituted complete with ecclesiastical courts as it had been before the civil wars.[36] The church now allied itself totally with the monarchy, preaching two political maxims: monarchy is divinely instituted and hereditary, and passive obedience is the primary responsibility of the subject should he be forced to face the dictates of an unjust monarch. Rebellion under any circumstances is sinful, the work of religious and political enthusiasts—irrational men driven by a distorted and perverse interpretation of Scripture and doctrine. On this interpretation of monarchical power and political obedience all churchmen agreed. There were other issues, however, over which deep discord existed between the Laudian or high-church faction and the moderates. These conflicts were rarely aired in public—neither side wished to give solace to the church's enemies.

The publication in 1662 of a tract by Simon Patrick, *A Brief Account of the New Sect of Latitude-Men,* provides one of the few sources for our understanding of the intellectual disagreements between the church's factions. Its author, at the time rector of St. Paul's, Covent Garden, had just been deprived of the mastership of Queens' College, Cambridge, by order of the king. Although Patrick had been duly elected by the fellows of the college, the king, apparently prompted by the leaders of the church, desired that the position go to Anthony Sparrow, a staunch Royalist. Throughout the Restoration, Patrick won little favor in hierarchical circles and in turn seems to have spurned such promotions as were passed his

---

[36] Robert Bosher, *The Making of the Restoration Settlement* (London, 1951), 143–218; Anne Whiteman, "The Re-Establishment of the Church of England, 1660–1663," *Transactions of the Royal Historical Society,* 5th ser., 5 (1955), 111–131.

way. The intellectual origin of Patrick's views, which provoked such hostility in some quarters, lies in his experiences at Cambridge during the civil wars and Interregnum.

The son of a prosperous mercer, young Patrick was intended for a career in business until, for reasons unknown, he chose to become a scholar and secured a place at Queens' College. His unpublished biography in the Cambridge University Library reveals that Benjamin Whichcote and Ralph Cudworth recommended him to Queens'. Patrick went up to Cambridge, therefore, already befriended by two of its leading churchmen. Once there Patrick grew close to Herbert Palmer, master of Queens', and John Smith, a member of Cudworth's philosophical circle. Palmer was reputed to be a Puritan, which in his case simply meant that he opposed Archbishop Laud and favored a loose form of ordination that would allow any company of ministers to ordain.[37] Perhaps it was as a result of Palmer's influence that Patrick accepted Presbyterian orders in 1648, an act which eventually made him uneasy. In 1654 he obtained private ordination according to the Anglican rite.

The chief formative influence on Patrick's thought was exercised by his close friend, the neo-Platonist John Smith. He confirmed young Patrick's suspicion that the doctrine of absolute predestination denies the true "nature of God, and his goodwill to mankind."[38] Patrick's denial of this main tenet of Calvinism, in which he apparently at one time had believed, allied him with the Cambridge Platonists and latitudinarians, who consistently attacked that doctrine.[39] Preaching against

[37] *DNB.* Palmer's manuscripts are in the Cambridge University Library.

[38] Biography of S. Patrick, MSS ADD 20, f. 14, U.L.C.; cf. S. Patrick, *The Autobiography of Simon Patrick, Bishop of Ely* (Oxford, 1839), 18.

[39] For a discussion of this change in Anglican thought away from predestinarian views, see John F. H. New, *Anglican and Puritan: The Basis of Their Opposition, 1558–1640* (London, 1964), 16–21. Cf. Robert Martin Krapp, *Liberal Anglicanism: 1636–1647* (Ridgefield, Conn., 1944).

strict Calvinism by the Platonic school was common in Cambridge during the late 1640s as the young Isaac Barrow's extensive notes on such sermons indicate.[40] Although willing to seek compromise with the Dissenters for the purpose of their comprehension into the church, low-churchmen attacked the doctrine of predestination as incompatible with their own sense of human free will and its role in the providential plan. Equally staunch was their anti-Catholicism, which in Patrick's case later earned him the censure of James II.[41] From the Cambridge school Patrick also received a rather simple version of the Platonic philosophy. He believed that "all this world below is but the image of the world above, and these corporeal things are but pictures (though pale indeed & dull), of things spiritual."[42] This sense of the transitory nature of the material order impelled Patrick's belief in the urgency of the Reformation, of the necessity of returning this order to its original, primitive purity in preparation for the consummation of the divine plan.

Speculations about the course of that plan and its time scale were common in Cambridge circles. Henry More's writings on the meaning of the Scriptural prophecies are well known. John Worthington, master of Jesus College and friend of Simon Patrick, also engaged in millenarian speculations, and later Simon Patrick lent his support to millenarian prophecies during the crucial summer of 1688.[43] The most famous of these Cambridge students, Isaac Newton, recorded in the early 1660s his belief in the approaching apocalypse.[44] The

[40] Notes on sermons by John Arrowsmith (1602–1659), Richard Vines (1600?–1656), Benjamin Whichcote, and Ralph Cudworth, among others, MS R.10.29, Trinity College, Cambridge.

[41] Patrick, *Autobiography*, 122–123.

[42] Patrick, *Aqua Genitalis: A Discourse Concerning Baptism* (London, 1659), A3.

[43] Patrick, *Autobiography*, 247. "Joseph" is a misprint. See John Worthington, *Miscellanies . . . Observations Concerning the Millennium . . .* , ed. E. Fowler (London, 1704); Jacob and Lockwood, 270.

[44] MSS ADD 3996, f. 101, U.L.C.

young churchmen tutored in Cambridge from 1645 to 1665 owed much of their religious conviction and sensibility to the influence of the Cambridge Platonists. Simon Patrick was but one of many who used their learning in pursuit of an ecclesiastical career.

The precise origin of Patrick's sympathy for the new experimental philosophy is unclear; possibly his brother John played a role. John Patrick posessed a genuine interest in mathematics and left several volumes of unpublished mathematical studies which have since been lost.[45] His unpublished notebook contains notes on Boyle's *New Experiments Physico-Mechanicall, touching the Spring of the Air* . . . (1660), and Simon used his brother's notebook to record his thoughts on popery and the need to preserve the Reformation.[46] In his sermons and writings Patrick adopts the practice common to the latitudinarians of using analogies to natural events or arguments from the new science to augment the Scriptural account of creation[47] or to express metaphorically the sense that the universe is the very tabernacle of the deity.[48]

Patrick's *A Brief Account of the New Sect of Latitude-Men*[49] deserves some attention. This apologia for the latitudinarians proposes to identify and justify the tenets of the new movement within the church. Patrick offers no precise definition of latitudinarianism—one would have been impossible, then or now. Rather, he describes this loose confederacy of churchmen according to their major assumptions about ecclesiastical disputes and about the relationship between philosophy and divinity. Latitudinarians stand between the "gaudiness of Rome" and "the squalid fluttery of fanatic con-

---

[45] MSS ADD 20, f. 117v, U.L.C.

[46] MSS ADD 77, U.L.C. See also MSS ADD 84(H) for J. Patrick's interest in natural philosophy during the 1650s.

[47] S. Patrick, *A Commentary upon the Historical Books of the Old Testament* (London, 1765), I, 1–23.

[48] Patrick, *Aqua Genitalis*, preface.

[49] Patrick, *A Brief Account* . . . (London, 1662; Los Angeles, Augustan Reprint Society, 1963, no. 100).

venticles" largely because of their experiences during the civil wars and Interregnum. They had "their education in the University, since the beginning of the unhappy troubles of this Kingdom, where they ascended to their preferments by the regular steps of election not much troubling themselves to enquire into the Titles of some of their Electours; they are such as are behind none of their neighbors either in Learning or good manners, and were so far from being sowred with the Leaven of the time they lived in that they were always looked upon with an evil eye by the successive usurping powers."[50] Trimming or the willingness displayed by low-churchmen to compromise with the powers that be is never presented by Patrick as an otherworldly retreat from political upheaval. It is rather a deliberate calculation of political behavior based on well-formulated suppositions. The moderates seek a settled liturgy, compatible with the interests of all reasonable Protestants, accept the Thirty-nine Articles of the church, yet continue to search Scripture and the church fathers for a true, primitive Christianity which not even the Reformation had successfully recovered. Reason guides their search, by which is meant simply a deductive faculty best exemplified in the writings of philosophers. The low-churchmen would revitalize the church and, as Patrick explains, return all things to their primitive pattern in conformity with the providential plan.

Providence instills order in the universe, "in the great automaton of the world," and modern philosophers, Tycho Brahe, Descartes, William Gilbert, Boyle, have made the greatest progress in the understanding of the Divine art. Their new philosophy, which Patrick broadly describes as "Platonick and Mechanick," will serve the church as no other. The adoption of the new philosophy by men willing to engage in political and ecclesiastical compromise will save the church and the Reformation against "the open violence of Atheisme, [and] . . . the secret treachery of Enthusiasm and Superstition."[51]

[50] Ibid., 5.
[51] Ibid., 19, 24.

Science will fulfill the Reformation and in the process defeat
the atheists and enthusiasts. By enthusiasts Patrick seems to
mean the radical sectaries, and by atheists he may be referring
to the Hobbists.

Similarities abound between Patrick's assessment of the new
philosophy and its use to the church and Sprat's discussion of
science. For direct contact or collaboration between Sprat and
Patrick no evidence exists. Nor need it. Sprat and Patrick
independently express views that were common in ecclesiastical and scientific circles. Never do they openly criticize the
existing ecclesiastical order, but criticism was certainly implied
in their advocacy of political moderation. Writing in the same
period, and not for public consumption, Isaac Barrow outlined
the many failings that prevented the church from making any
headway in the comprehension of Dissenters. Its services were
"long and tedious; which distasteth men of business and dispatch. It might suit cloisters (whence much of it came) rather
than congregations of tradesmen and merchants," and, not
least of all, the church spurns men "of good learning and good
life, of modest and peaceable dispositions, moderate in their
judgments."[52] True to Barrow's complaint, only a few moderates attained bishoprics in the church during the Restoration. Equally interesting, in light of Barrow's complaint about
how inappropriate existing church services appear to "men of
business and dispatch," is Barrow's own family background,
so similar to that of other low-churchmen. Like Patrick, Wilkins, and Tillotson, he came from a family engaged in trade
and commerce. They rose in the church by dint of hard effort.
As masterful preachers they sought power through their
pulpits, using simple and direct language to explain their social
and political philosophy.

Occasionally public quarrels erupted between the church's
factions. Efforts were made in the 1660s to deprive Henry

[52] *BTW*, IX, 578–580; on the question of dating see Osmond, 76–77.

More and Ralph Cudworth of their college livings.[53] In July 1669 at the opening of the Sheldonian Theatre, Robert South preached against "fanaticks," comprehension, the Royal Society, and the new philosophy, themes he labored in other sermons, eventually published.[54] Possibly in response to his baiting, Edward Fowler defended the moderates the following year in a treatise entitled *The Principles and Practices, of Certain Moderate Divines . . . Called Latitudinarians. . . .* Against the accusations of Dissenters who find the low-church compromise with church ritual and practices to be odious and self-serving, Fowler justifies his faction not by defending church rites and doctrines, but by reference to the authority of Calvin, Beza, and others, Protestant apologists who supposedly preached the virtues of ecclesiastical peace and conformity for pragmatic reasons. The latitudinarians, as Fowler sees them, are sensitive to every issue confronting the church, so much so that their very style of preaching has been accommodated to these issues.[55] Fowler lays emphasis on pulpit oratory that will present a reasonable version of Christianity in straightforward language. The latitudinarian effort to inculcate a natural religion that would bridge the differences among Protestants produced a new rhetoric, best exemplified in the sermons of Tillotson, Barrow,[56] and Wilkins. They eschewed florid metaphors and complex analogies, not simply as a matter of sensibility or aesthetics, but because they wished to inculcate the principles of natural religion to listeners en-

[53] Marjorie H. Nicolson, "Christ's College and the Latitude-Men," *Modern Philology*, 18 (1929), 50–51.

[54] H. Oldenburg, *The Correspondence*, trans. and ed. A. R. Hall and M. B. Hall (Madison, 1969), VI, 129; Boyle, *Works*, VI, 458–459. See also Robert South, *Sermons Preached upon Several Occasions* (Oxford, 1823), I, 373–374, 58–59.

[55] Fowler, *The Principles and Practices . . .* (London, 1670), 25, 91, 40–41, 103.

[56] See Irene Simon, *Three Restoration Divines: Barrow, South, Tillotson, Selected Sermons* (Paris, 1967).

gaged in business or trade, many of whom had been steeped in the tradition of Puritan sermonizing. The latitudinarians were adopting a style of preaching made common by the Dissenters and praised by Sprat in his *History* for its usefulness in describing natural phenomena and the real world of sense perception.

Although Fowler may have seen value in the Puritan style of preaching, he had no use for Calvinist doctrines. Absolute election is untenable, he asserts, for "none . . . are damned but those that wilfully refuse to co-operate with that grace of God, and will not act in some suitableness to that power they have received."[57] The latitudinarians would make men responsible for their actions because they feared that in the tenets of extreme Calvinism radicals would find, as they had found during the Revolution, justification for abrogating individual responsibility for one's plight and laying blame on established institutions.[58] Furthermore, within strict Calvinism lay the possibility of justifying Anabaptist philosophies, which were equally dangerous to any ecclesiastical institution claiming the right to mediate between God and man. Fowler makes it clear that the latitudinarians are firm supporters of episcopal government and of the supremacy of the civil magistrate in both civil and "sacred" affairs.[59]

Despite Fowler's defense, high-church invective against the latitudinarians continued. In 1675, during a sermon delivered before king and court, John Standish chastised "those false Apostles" who "would supplant Christian Religion with Natural Theologie; and turn the Grace of God into a wanton Notion of Morality; . . . [and] impiously deny both the Lord . . . and his Holy Spirit . . . , making Reason, Reason, Reason, their only Trinity."[60] Simon Patrick responded, ac-

---

[57] Fowler, 228–229.
[58] Hill, *World*, 130–135.
[59] Fowler, 324–335.
[60] Standish, *A Sermon Preached before the King at Whitehall, September the 26th 1675* (London, 1676), 24–25.

cused Standish of gratifying the church's enemies, and demanded that he name the false apostles, which Standish
promptly did.[61] He accused them of Arminianism and Presbyterianism, but for all the anger of Standish's remarks, his
charges were vague and unformulated. During the next two
years John Warly added fuel to the controversy and labeled
the latitudinarians as apostates.

To negotiate with the Dissenters is to countenance theories
that undermine kingly power and to invest it in the hands of
magistrates. Warly uses low-church efforts at comprehension
to accuse the latitudinarians of being Calvinists—a charge
without doctrinal foundation—and accuses the latitudinarians
of being the new instigators of civil war. Finally he proposes
that they are preaching a new religion, that the moderates
"imagine there were to be virtuosos in Religion as well as
Philosophy, and that this age made new discoveries of Doctrines as the Astronomers have of the Stars, and that new
Creeds in Divinity are as necessary as new Systems of the
World."[62] The use of science to further religious goals naturally
incited churchmen who remembered the Puritan advocacy of
Baconian science in the 1640s.[63] Some churchmen simply dis-
trusted any intellectual novelty, associating it with civil
disorder and religious dissension. Warly and the circle for
which he spoke[64] eschewed any religious certitude based

[61] Patropolis, *An Earnest Request* . . . (London, 1676). In all probability, the author is Patrick; see *Autobiography*, 79; John Standish,
The Truth Unvailed, In Behalf of the Church of England (London,
1676), naming S. Parker, Sherlock, probably William Hickring, a Dr.
Hall, among others.

[62] Warly, *The Reasoning Apostate: or Modern Latitude-Man Consider'd* . . . (London, 1677), 86–87. See also James Duport, *Musae
Subsecivae, seu Poetica Stromata* (Cambridge, 1676), 58.

[63] Christopher Hill, *Intellectual Origins of the English Revolution*
(Oxford, 1965), chap. 3.

[64] His treatise, *The Natural Fanatick, or Reason Consider'd in Its
Extravagance in Religion, and (in some Treatises), Usurping the Authority of the Church and Councils* (London, 1676), was encouraged
by Heneage Finch, earl of Winchilsea, a prominent church patron, a

largely on convictions derived from rational disputation. Their position resembles fideism and they doubted that skepticism could be cured by the use of arguments derived from the natural order.[65] Consequently Warly found the Cartesian version of the mechanical philosophy acceptable, since for him proof of divine efficacy did not hang on the operations of matter and motion, and the danger of materialism hardly mattered.[66] Both philosophically and politically, high-churchmen rejected the methods by which the moderates sought to secure the church's power. Some contemporaries for whom Warly may have spoken saw latitudinarianism as so dangerously radical as to constitute (in effect) a new religion.

The charge is not without substance. The latitudinarians devised a natural religion comprehensive enough to override doctrinal differences and so broad in its application as to include behavior once labeled simply as unchristian. Through a vast sermonizing effort their thought reached large congregations mainly in London, whose involvement in business, politics, or trade made them prosperous and, in some cases, the leaders of post-Restoration society. Every sermon had a particular audience, one guide to the composition of which is the parish in which the preacher spoke. Tillotson preached almost every Tuesday at St. Lawrence Jewry, having been chosen as its lecturer by a plurality vote of churchwardens and about forty parishioners.[67] There he addressed parishioners as well as officials from the Guildhall. The baptismal records of the parish indicate that an exceptionally large proportion of the parish engaged in trade either as small craftsmen or merchants. Latitudinarian sermons commonly were given in other

---

complicated business, since at times the Finches were great patrons of the latitudinarians. The shifting political alliances of the Restoration and their effect on the church call for a careful study.

[65] Ibid., 25, 33.

[66] Ibid., 42.

[67] "Vestry Book of St. Lawrence Jewry," MS 2590/1, f. 545, Guildhall Library, London.

prosperous parishes, St. Margaret, Westminster, Lincoln's Inn,[68] St. Paul's, Covent Garden, St. Andrew's, Holborn, and St. Martin's-in-the-Fields, reputed to be the wealthiest parish in the city. The composition of the parish is not, however, an infallible guide to the makeup of any particular congregation, since it was fashionable to travel about the town in search of one's favorite preacher, or simply to see and be seen. Outside of London, Isaac Barrow usually preached in Cambridge, either at college chapel or at Great St. Mary's, where John Moore also delivered an important set of unpublished sermons against Hobbes which will be discussed shortly.

From the sermons and treatises of Isaac Barrow, John Tillotson, John Moore, and John Wilkins, it is possible to cull a synthetic view of latitudinarian natural religion and at the same time to define the justifying and supportive role played in it by scientific knowledge and the new philosophy. If this method ignores individual differences, it is because they do not appear so significant as to have provoked controversy or public disagreement among the latitudinarians for whom there existed surprising agreement. They addressed themselves primarily to the public actions of the Christian; they seldom discussed the nature of inner spirituality or private communications with the creator. Public piety interested the latitudinarians, particularly as it influenced the decisions taken by the prosperous in their pursuit of virtue and interest.

Much of what can now be said about the social theories of seventeenth-century churchmen relies heavily on the excellent work of Richard Schlatter, and this study is no exception. The views of the latitudinarians discussed here, particularly about

[68] A. W. Hughes Clarke, ed., *The Register of St. Lawrence Jewry 1538–1676*, I, 65–75, in vols. 70–71 of publications of *The Harleian Society* (London, 1940); see also D. V. Glass, "Socio-economic Status and Occupations in the City of London at the End of the Seventeenth Century," in A. E. J. Hollaender and William Kellaway, *Studies in London History* (London, 1969), 373–392. See David D. Brown, "The Text of John Tillotson's Sermons," *The Library*, 5th ser., 13 (1958), 29.

work and charity, have received attention from Schlatter, and my own study supplements his account by enlarging his list of Anglicans and by offering a somewhat different perspective on their motivation. My evidence suggests that certain Anglicans were neither more nor less keen on promoting industry and trade than were the nonconforming preachers.

Beyond that, however, the crucially important role played by scientific knowledge and natural philosophy must be emphasized. Science acted as an anchor for social theories and provided an underpinning that synchronized the operation of nature with an economic and social order increasingly determined by capitalistic forms of production. To my mind, the most historically significant contribution of the latitudinarians lies in their ability to synthesize the operations of a market society and the workings of nature in such a way as to render the market society natural. The latitudinarians grafted the new philosophy onto their social ideology, integrating both into English thought precisely at a time when modern and capitalistic forms of economic life and social relations were gaining ascendancy. Their synthesis survived without serious questioning, except by deists, freethinkers, and atheists, because it gratified the beneficiaries of that new order. Natural religion made the actions of the prosperous compatible with Christian virtue and with the very mechanism of the universe.

According to their preachers, the prosperous benefited because they worked, and success in this world as well as in the next rested not on any imagined predestination but on an act of individual will. The latitudinarians, by and large, would have nothing of Calvin's doctrine because it eliminated the necessity of striving and undermined the virtue of self-help.[69] Their belief that man must win salvation from a just and providential deity who would never ruthlessly close the contest before it had begun complemented their view of work as

[69] *BTW*, V, 140; IV, 290–291. Tillotson, reflecting his Presbyterian background, was prepared to accept a loose form of predestination while emphasizing the existence of free will; *TW*, II, 495.

the prerequisite for prosperity, or, if one were unlucky, for the privilege of receiving charity.

The latitudinarian rejection of predestination appears paradoxical and even surprising in contrast to their militant commitment to the Reformation. Almost certainly their return to what was, in effect, a doctrine of good works occurred in response to the uses to which predestination was put by the radicals in the late 1640s and 1650s. The doctrine of predestination, like that of providence, possessed protean ideological qualities. If the radicals could use it to support their ends, then latitudinarians could effectively abandon the doctrine, but not acknowledge their retreat from one of the central doctrines of the original reformers. The latitudinarians thought ideologically, rather than with constant attention to theological subtlety.

Basing social status and economic inequality on the work ethic, the latitudinarians with their characteristic realism recognized and indeed encouraged social inequality. At times the latitudinarians justified it simply by reference to the work ethic, yet more commonly they lapsed into thinking that inequality was in the natural order of society.[70] Yet the origin of private property bothered them, not because of injustices, but because Hobbes had offered an explanation that relied not upon hard work or natural law but solely upon the mechanisms of force and power.

Among the many enemies against whom the latitudinarians preached (Catholicism was a favored topic) none disturbed them more deeply than Hobbes. They framed their objections to his philosophy almost entirely by reference to the new philosophy as explained by Boyle, Wilkins, and the Cambridge Platonists.[71] At the heart of their criticism lay the

[70] Richard Schlatter, *The Social Ideas of Religious Leaders, 1660–1688* (Oxford, 1940), 89–105.

[71] There was a vast industry devoted to attacking Hobbes. See S. Mintz, *The Hunting of Leviathan* (Cambridge, 1962); for bibliography, Q. Skinner, "Thomas Hobbes and the Nature of the Early Royal Society," *Historical Journal*, 12 (1969), 217–239.

belief that Hobbes and Hobbism constituted a grave moral threat to the fabric of social relations and hence to order and stability. The latitudinarians persistently repeated one theme: Hobbes's philosophy encourages rapacious self-interest, the pursuit of profit and status by ungodly men whose success galls and even imperils men of virtue. Indeed the latitudinarian analysis of Hobbes, in effect, to see him as a critic of social relations in a market society, surprisingly resembles the approach taken to him by modern commentators, most notably by C. B. Macpherson.[72] The latitudinarians presumed, perhaps without sufficient evidence, that the market society had arrived. More than any other segment of the church, they were sympathetic to "men of business and dispatch." It should not be imagined, however, that the latitudinarians attempted simply or primarily to offer a defense of bourgeois values and ethics. The recent interpretation of Hobbes as, in effect, accepting bourgeois morality and as justifying the bourgeois state should not be applied to the latitudinarians, at least not without important qualifications. From the latitudinarian perspective, why defend a society that had already come into being, when the main task lay in seeing to it that the forces of greed and disorder did not get out of hand entirely, foment disorder, destroy Anglican moral and social leadership, and defy the providential plan.

A complex relationship existed between the latitudinarians and their bête noire, Hobbes. Both recognized the emergence of a new form of social relations which required new understanding and insight. Churchmen believed that Hobbes had spoken to his society as had no other philosopher and in so doing had encouraged its worst excesses. He was for them a symbol of their gravest misgivings about the market society.

[72] Macpherson, *The Political Theory of Possessive Individualism, Hobbes to Locke*; for a critique see K. Thomas, "The Social Origins of Hobbes' Political Thought," in *Hobbes Studies*, ed. K. Brown (Oxford, 1965); M. C. Jacob, "The Church and the Formulation of the Newtonian World-view," *Journal of European Studies*, 1 (1971), 132.

The latitudinarians wished to address themselves to that same society, and they could only do so by accepting some of its basic values and assumptions and therefore, in a sense, becoming defenders of capitalism. More precisely, however, they proposed what can best be called a Christianized capitalism, an ethic for self-interest resting upon the providential order in the world political and natural. The latitudinarians differed profoundly from Hobbes in presuming that market relations must be Christianized and that their natural religion best suited the task. At every opportunity churchmen attacked Hobbes, yet the similarity of their concerns over restraint and control in a market society meant that occasionally they and their enemy would betray similar insights into the origins of private property or into bourgeois psychology.

Most churchmen justified the existence of private property by reference to the laws of nature; in the state of nature property would be sacrosanct.[73] Yet occasionally they saw the dangers inherent in that approach. To maintain that private ownership was simply in the order of nature might induce complacency and ingratitude toward the providential deity, without whom "all thy projects were vain, all thy labours . . . fruitless."[74] Alternatively, to maintain property rights as inviolable would justify the current distribution of property and wealth, which many in the church regarded as inadequate to its needs. In response to this dilemma, churchmen occasionally claimed that ownership is based on expediency and not on inherent and natural right—a position dangerously close to that of Hobbes.[75] Like him the latitudinarians recognized that property equaled power;[76] therefore the allocation of property to the poor presented certain difficulties. The problem was

[73] Schlatter, 89–95.
[74] *BTW*, I, 32–33; John Moore, "Sermons," MS Dd. 14.9, f. 39, U.L.C., "sin, which first put men on distinguishing into proprietys."
[75] Schlatter, 102–103.
[76] For a plain-spoken statement of the assumption, see H. Burton, *A Second Volume of Discourses* (London, 1685), 337.

vexed by the apparent growth in the number of poor in London and to a lesser extent elsewhere in the countryside.[77]
Charity was, of course, a Christian virtue, but the latitudinarians would have it applied with discrimination. Stillingfleet
believed that those who voluntarily chose poverty out of
laziness deserved to starve or should be put in workhouses.[78]
Tillotson preached that the poor who came to that condition
through no obvious fault of their own bore witness to God's
providence because they provide an occasion for the practice
of charity by the rich. Indeed the practice of charity, aside
from being a Christian duty, enables the rich to insure their
wealth against those who would deprive them of it. There was
just so much charity to go around, and the latitudinarians
would have it distributed to the most deserving—to those who
have honest callings and are in difficulty or to those who by
force of circumstances have fallen from riches.[79]

The latitudinarian assessment of the relationship between
virtue and profit entailed far more than a call for the discrete
distribution of surplus wealth. The natural religion they articulated was intended to suit the needs of the prosperous, or
to be more precise, the strivings of those who would seek profit
and remain virtuous. According to Barrow, everyone seeks
profit, that "great mistress," and given the fact of human
covetousness, piety is useful because it provides security in this
world or the next.[80] If harmony and stability are to prevail, a
natural religion that condones human needs for power and
profit is essential. Restoration churchmen were obsessed with
the problem of maintaining social and political stability[81] and

[77] Hill, *World*, chap. 3; J. Sharpe, *A Sermon Preached before the
. . . Lord Mayor . . .* (London, 1680), 27.
[78] Stillingfleet, *Protestant Charity, A Sermon Preached at St. Sepulchres Church, on Tuesday in Easter Week* (London, 1681), 13.
[79] *TW*, II, 474, 593–597; Sharpe, 21; T. Tenison, *A Sermon Concerning Discretion in Giving Alms. Preached at St. Sepulchres Church
. . . , 1681* (London, 1688). Cf. Schlatter, 131–132.
[80] *BTW*, I, 175–177, 215.
[81] R. Harvey, "The Problem of Social-Political Obligation for the

sought it in a creed that would neutralize the sources of discontent by channeling private interest into the service of public interest. Wilkins and Tillotson explicitly argue that their natural religion is a public religion[82] that condones and fosters the pursuit of self-interest. Only God is not self-interested, and as a result he aids our interests[83] as does religion itself. It conduces to profit and abets our "sober self-love,"[84] precisely because natural religion instills certain public virtues. Under the influence of latitudinarian natural religion princes will be just and subjects will obey, relations between inferiors and superiors become benign, and, most important, men will be diligent in their calling, indeed they will be impelled by religion to pursue their interests and presumably to attain them. So probable is the success of the virtuous that their prosperity is an even higher sign of God's providence than is the order inherent in nature.[85] Only true industry constitutes the proper pursuit of self-interest; indeed Scripture recommends it as both just and necessary; covetousness and greed are actually antithetical to the fulfillment of desire.[86] To John Moore the rapaciousness supposedly advocated by Hobbes demanded the constant guarding of one's life, and this meant the cessation of interest and presumably the curtailment of profit and power.[87]

Similarly, the latitudinarians were convinced that Hobbism provided the major obstacle to the acceptance of natural religion and worked ultimately against the preservation of society

Church of England in the 17th Century," *Church History*, 49 (1971), 156–169; see also George Williamson, "The Restoration Revolt against Enthusiasm," *Studies in Philology*, 30 (1935), 571–604.

[82] *TW*, II, 300–301; J. Wilkins, *Of the Principles and Duties of Natural Religion*, ed. Henry G. Van Leeuwen (New York, 1969), 168.

[83] *BTW*, I, 363; VII, 158–159; Burton, 335.

[84] *BTW*, I, 177; IV, 141; Wilkins, 304–306, 312; *TW*, III, 53–55.

[85] Wilkins, 174–175.

[86] *TW*, II, 224–225. Cf. N. Sykes, "The Sermons of John Tillotson," *Theology*, 58 (1955), 300.

[87] John Moore, MS Dd. 14.14, f. 9, U.L.C.

and government.[88] Because the latitudinarians shared a similar view of human nature with their rival,[89] they were convinced of his potency. On occasion they too saw fear, particularly of eternal punishment, as the only restraint, although in its place they would substitute a religion that catered to the desire for profit and power while at the same time restraining it. By the cultivation of socially useful virtues—diligence in one's calling, calculated charity toward inferiors, obedience in familial and political relations, practicality among the rich and leisured, and, above all, resignation to one's material circumstances whatever they may be—the latitudinarians would bend private interest to serve the public good. Desire could never be satisfied; only moderation in the pursuit of interest could control desire and thereby ensure social stability. The basis for latitudinarian moderation, whether in politics or religion, lay in the belief that self-interest could be tamed and yet fulfilled, that interest had its rewards, both temporal and eternal.[90]

Making moderation work required a particular notion of the deity and his relationship to creation. The belief in a providential God is basic Christian theology, but like every broad and complex concept it has protean qualities and is capable of various interpretations. During the Reformation each side believed itself to be part of the providential plan and believed its actions to be merely a fulfillment of preordained and divine intentions. In 1588 the English rejoiced in the Protestant wind because of their belief in providence; at the same time the Spanish resigned themselves to defeat by pondering the mysterious ways of God. In the sixteenth century providence became an integral part of Protestant polemics, no less in England than on the Continent. Within English Prot-

[88] Ibid. T. Tenison, *A Sermon against Self-Love . . . 1689* (London, 1689), 16–17.

[89] *BTW*, VI, 430–433; Wilkins, 167–168; at times their thinking on this point is contradictory, *BTW*, I, 167. Tillotson was accused of preaching Hobbist doctrines and changed his text to accommodate his accusers; Brown, "Text of John Tillotson's Sermons," 18–36.

[90] *BTW*, VI, 4; II, 303–304; *TW*, II, 231–234, 242; Wilkins, 87.

estantism, providentialism came increasingly to be associated with the Puritan sectaries and, in its extreme but logical form, the doctrine served as the basis for millenarian speculation. Belief in providence entailed belief in a divine historical plan revealed, if at times cryptically, in Scripture. Those radicals who wanted to create a new social order argued that the established church represented antichrist and occasionally so did the king and the nobility. The defeat of antichrist had been foretold in Scripture; it was part of the providential plan. The task of the radical sectaries entailed the accomplishment of that plan as quickly as had been intended. To other Protestants their failure meant simply that they had misread the plan, and the final collapse of Puritan ascendancy at the Restoration convinced churchmen that in the operation of divine providence the church played the crucial role.

The more militant and anti-Catholic churchmen, and certainly the latitudinarians belonged to that camp, now assumed that the fulfillment of the Reformation depended solely upon the ascendancy of English Protestantism as represented by the church. First, there must be the unification of English Protestantism, that is, the comprehension of the Dissenters, second, a series of ententes with Continental churches and simultaneously the propagation of the gospel in foreign lands. As the strength of international Protestantism grew, the Roman church—the antichrist—would wither and eventually, either through spiritual or military conquest, the beast would be slain, and the fourth and final empire, that of Rome, would fall. History would end at the second coming, and the saints led by the English church would reign triumphantly in the millennial paradise.[91] Certain minor details would have to be worked out en route: the Jews and Mahometans would be converted, and those who resisted any form of religious worship, the immoral and atheistical, would be shown the error of their ways. The ascendancy of the latter group disturbed churchmen. It was as if the meaning of Scripture was par-

[91] See Chapter 3.

ticularly cryptic on that point, and their growing strength had not been properly and clearly foretold.

The natural religion of the latitudinarians was intended as the vehicle by which the historical dimension of the providential plan would be accomplished most easily and efficiently. A broad and liberal Protestantism, only marginally concerned with ritual and doctrine,[92] would facilitate the comprehension of Dissenters, at the same time convincing godless and worldly men that religion lay in their best interests. The instinct for profit and power became for the latitudinarians a part of the providential order. Had not providence established the difference between rich and poor as an occasion for practicing virtue? Could God not reverse a man's fortune at will?[93] The precarious nature of market forces spring from the very nature of the providential order. Beneath this only apparent uncertainty lay a plan that instilled order and maintained harmony despite the greed and self-love of men. According to Tillotson, every aspect of human endeavor, government, trade, and commerce,[94] bears witness to the providential plan. When men imagine that they can manage their affairs without the guidance of providence, injustice results. Monopolies and fixed prices disrupt the providential order; market forces are ultimately more just, for they rely on natural forces rather than on the greed of unbridled self-interest.[95] But interest brought into conformity with the providential plan will reap a rich harvest—industry coupled with charity and restraint produces prosperity—and private prosperity means public prosperity.[96]

Social relations in a market society, according to the latitu-

[92] *TW*, II, 311–312; Stillingfleet, *A Sermon . . . 1690/1* (London, 1691), 22.
[93] *TW*, II, 474; Burton, 335.
[94] *TW*, II, 554–555.
[95] Moore, MS Dd. 14.9, f. 32, U.L.C.; J Tillotson, "On the Rule of Equity . . . 1661," in Paul A. Welsby, ed., *Sermons and Society* (Baltimore, 1970), 136.
[96] *BTW*, I, 24; II, 234–235.

dinarians, were governed by providence and could be fitted into the providential plan. Their concept of providential intervention divorced it, by and large, from direct confrontations with antichrist—although the latitudinarians still allowed for such events—and from the radical interpretation of the providential plan which justified revolutionary change. Instead providence oversaw the practitioners of natural religion as they pursued their interests, prospered, and thereby served God and church. The latitudinarians were able to imagine the hand of God at work invisibly and indirectly, in social and economic relations—because their concept of the natural order rested on the new mechanical philosophy. Natural philosophy told churchmen like Wilkins, Barrow, Tillotson, and others that the world natural operated according to the laws governing matter and motion but only through the assistance of spiritual forces, in other words, of providence. The sophisticated version of providential action in the "world politick" and the world natural preached by the latitudinarians relied on secondary causes, laws of nature or of social behavior, for the operation of both realms.[97] John Moore, arguing against Hobbes, preached that only the evidence of laws of nature operated by spiritual forces in nature would give the lie to Hobbes's social philosophy, and Moore believed that knowledge of those laws came from mathematics and metaphysics, or more generally, from science and natural philosophy.[98]

The social philosophy of the latitudinarians could rest on their understanding of the natural order, on the discoveries of the new science,[99] because they believed that the worlds natural and political were interrelated. One variation or other of that belief was commonplace in early modern English thought.[100] Generally it is described as the microcosm-

[97] *TW*, I, 330; II, 572; Moore, MS Dd. 14.9, f. 80, U.L.C.
[98] Moore, MS Dd. 14.15, ff. 8–9, U.L.C.
[99] Richard Schlatter appears to see this relationship, but its examination was outside the scope of his study (201, 224).
[100] On a related theme, see M. Macklem, *The Anatomy of the*

macrocosm theory, but in this regard I want to distinguish latitudinarian thought from the simple rendering of that theory. The latitudinarians generally did not dwell upon direct divine intervention. Exceptional providences were allowed, of course, such as the Protestant wind of 1688. But they did not believe that thunderstorms or earthquakes need necessarily denote moral disorder as once they might have. In general, disruptive natural events were governed by laws or secondary causes that rendered them neutral or without moral connotations. Yet at the same time the latitudinarians freely acknowledged that extraordinary natural events could be signs from God.[101] The moral principles of natural religion were derived by churchmen, however, from the very order and regularity of nature, from the operation of general providence. The design and harmony in the material order, imposed by spiritual forces, provided a model or guide to show how social and political relations should work if Christians were to fulfill the providential plan and still compete with one another.

The operations of the natural world as explained by the new mechanical philosophy provided churchmen, both before and after Newton, with a model, which by analogy would enforce stability and harmony in the "world politick." Both realms were entwined in the providential plan; both human and natural history would terminate simultaneously at an appointed time. Occasionally, in extraordinary times of instability, such as the events of 1688–1689 which will receive discussion in Chapter 3, an earthquake could forebode an immediate disaster in the political order or a comet could crash into the earth as retribution for human sinfulness. Churchmen such as

---

World: Relations between Natural and Moral Law from Donne to Pope (Minneapolis, Minn., 1958).

[101] This was a vexed problem in eighteenth-century thought. I am simply stating the general latitudinarian position. See Jacob Viner, *The Role of Providence in the Social Order: An Essay in Intellectual History* (Philadelphia, 1972), 16–18.

Newton did imagine that such catastrophes could occur,[102] but in general terms it was the constant and regular pattern of nature that was important.

Barrow used the operation of nature by providence to argue that the existing political authority can never be overthrown: "God's providence is the only sure ground of our confidence or hope for the preservation of church and state, or for the resitution of things into a stable quiet." Because "letting the world alone to move on its own hinges" under the guidance of providence ensures stability in nature, we should presume that providence will guide legitimate authority in the management of political affairs.[103] One of Barrow's sermons provides the most succinct statement of the crucial relationship that the latitudinarians imagined between the worlds political and natural: "As in the world natural, the parts thereof are so fitted in varieties of size, of quality, of aptitude to motion, that all may stick together, (excluding chasms and vacuities,) and all co-operate incessantly to the preservation of that common union and harmony which was there intended; so in the world political we observe various propensions and attitudes disposing men to collection and coherence and co-operation in society."[104] Social harmony relies on the providential plan, and nature reveals that plan, in the design of species, the regularity of seasons, the regular cycle of the universe. The design and restraint evident in the laws of nature, as John Moore understood them, allow men to derive a morality of restraint in their social relations which is the only sure foundation of prosperity and government.[105]

The latitudinarians proposed that because the atoms render

[102] See David Kubrin, "Newton and the Cyclical Cosmos: Providence and the Mechanical Philosophy," *Journal of the History of Ideas*, 28 (1967), 325–346, and J. E. McGuire, "Transmutation and Immutability: Newton's Doctrine of Physical Qualities," *Ambix*, 14 (1967), 69–95.

[103] *BTW*, I, 435.

[104] *BTW*, V, 231–232; cf. Wilkins, 158.

[105] *BTW*, V, 231–235; Moore, MS Dd. 14.15, ff. 8–8ᵛ, U.L.C.

matter into shapes and sizes, not by chance but by an inherent
tendency for cooperation and coherence, men should likewise
exist in relative social harmony. Industry and work pervade
nature and ensure its stability, and this the latitudinarians took
for further proof of the validity of the work ethic. Similarly
every plant, every animal, every planet has a place, and men,
if they follow their callings, have a preordained and unalterable place in society. The calm and quietude of the heavenly
bodies resemble the calm desired in the commonwealth.[106]
Tillotson informed his congregation that science proclaims
and demonstrates the inherent order[107] within nature, and
thereby ensures the interests of those who benefit from order
and stability.

But not all science and natural philosophy serve to promote
social order and stability. Churchmen often attacked the recently revived atomism of Epicurus. Although Christian
natural philosophers from mid-century onward came to accept the atomic structure of matter—indeed it is at the foundation of Newton's system of the world—they did so only after
they had discarded Epicurean materialism and determinism.
Stillingfleet and Tillotson, and later Richard Bentley, argued
that the atomism of Epicurus denies providence and consequently is dangerous to the social and political order.[108]
Atomism was only acceptable to the latitudinarians if the
principles of motion by which the atoms are formed into material substances are rendered external to them and therefore
controlled by spiritual forces.[109]

Similarly the latitudinarians such as Barrow and Tillotson rejected Cartesianism because it allowed for a mechanical explanation of motion devoid of spiritual forces and, therefore, devoid of divine providence. Previous commentators on

[106] *BTW*, V, 204; III, 389–391, 410–413; VII, 52; II, 204–205.
[107] *TW*, II, 551–552.
[108] Stillingfleet, *Origines Sacrae*, 378–379; *TW*, II, 558–559; see
Chapter 5.
[109] See Kargon, *Atomism*.

the reception of Descartes in England have stressed the religious objections of English natural philosophers to his system. But what did religion mean to these churchmen? It was much more than personal piety devoid of ideological content. Providence, spiritual forces, active principles, lifeless matter, space as a near-extension of God's being were concepts by which the radicals, the materialists, and the Epicureans could be defeated. And with their defeat came prosperity and stability; the church's power would be guaranteed. Those very concepts first articulated during the Restoration became the metaphysical underpinning of Newton's system of the world, and without those philosophical assumptions the Newtonian achievement might never have occurred.[110]

For Restoration churchmen natural philosophy served ideological and political needs. We should hardly be surprised, therefore, to find their successors in the church's championing, for similar reasons, Newton's system of the world. In the *Principia* (1687) and later in the Queries to the *Optice* (1706) Newton offered mathematical formulae and scientific observations that worked admirably to explain the order of the universe, and he based his understanding of how universal gravitation operated on the same natural philosophy that churchmen had found most appealing as a weapon in their struggle with the radicals and materialists.

Only a natural philosophy such as Newton's that embodied nonmechanical assumptions about nature was compatible with the social philosophy of the latitudinarians. Matter had to be

[110] *TW*, III, 122; Stillingfleet gradually abandoned Descartes. The flirtation of the Cambridge Platonists with the Cartesian philosophy has been amply demonstrated. See n. 8, above. I am not implying here, or throughout this essay, that Newton's own rejection of Descartes can be explained simply by reference to social and religious considerations. M. Nicolson, "The Early Stages of Cartesianism in England," *Studies in Philology*, 26 (1929), 356–374; Sterling P. Lamprecht, "The Rôle of Descartes in Seventeenth Century England," Columbia University, Department of Philosophy, ed., *Studies in the History of Ideas*, 3 (1935), 181–242.

dead and lifeless—passive[111]—only then could providence be said to operate and spiritual forces be made dominant in the natural order and in the affairs of men. If matter moved by its own inherent force, God would be rendered useless and men would pursue their interests unimpeded. If matter was infinite and active and space merely a relative notion, as Descartes would have it, then both space and matter would be eternal and independent of God.[112] The result of such a natural order would be to sanction the Hobbesian world where the greed of the great and the desires of the poor wreak havoc and deprive the virtuous of their just rewards. Barrow said that in Hobbes's world the church would founder and its sectarian enemies, like "prodigious meteors," would collide and burn until they had in the course of their own destruction implanted "anarchy, emulation, and strife." Were such a point reached Tillotson argued that a merciful but vengeful creator would destroy forever an evil and ungrateful world.[113]

To prevent such a holocaust, or more precisely to ensure its arrival not as retribution for the failings of the saints but as a fulfillment of the providential plan, the latitudinarians enlisted the new natural philosophy. They understood that philosophy largely in terms of its social implications. The conclusion appears inescapable: social and political pressures in turn played a crucial role in shaping the new science. The thinking of Robert Boyle[114] and the sermons of Isaac Barrow provide substantial evidence that they formulated their understanding of the world natural at least partially in response to

[111] *BTW*, VII, 133–137; Moore, MS Dd. 14.9, f. 80ᵛ, U.L.C.
[112] I. Barrow, *The Usefulness of Mathematical Learning . . . Being Mathematical Lectures Read in the Public Schools . . .* (London, 1734), lecture 10. For Newton's debt see E. W. Strong, "Barrow and Newton," *Journal of the History of Philosophy*, 8 (1970), 155–172.
[113] *BTW*, IV, 20, 28; *TW*, II, 575–576; III, 170–172, 247–248; II, 641.
[114] See J. R. Jacob, *Boyle*. Cf. his "The Ideological Origins of Robert Boyle's Natural Philosophy," *Journal of European Studies*, 2 (1972), 1–21.

a political and social order that required a new natural model. The development and acceptance of the new mechanical philosophy must be understood with constant reference to the social context within which that intellectual achievement occurred. From the time of Boyle to that of Newton, England experienced a political and social revolution. The subsequent emergence of new economic and social patterns best described as capitalistic gave rise to the need, at least as latitudinarians saw it, for a new natural religion. In this context the new philosophy developed as a response to changing social realities just as it also synthesized the information gathered from experimentation or mathematical reasoning.

Churchmen realized that the implementation in society of this new natural model would require considerable effort on their part. Although they argued for order and harmony and enlisted the workings of the entire universe as supreme justification, clearly they were worried. They knew that prosperity and stability counted for everything, yet the wrong people appeared to be prospering. The church had to prosper or be "ill-respected,"[115] but by the late Restoration its fortunes, certainly political and quite probably economic, were lagging if not faltering. Although we possess no adequate study of church revenues in this later period, churchmen believed them to be falling.[116] Bishop Isaac Barrow, uncle of the famous Barrow, took such drastic steps to recover church lands on the Isle of Man that his actions might be best described in latitudinarian terms as Hobbist.[117] By the reign of Anne the lower clergy complained bitterly of their poverty. Whenever churchmen address themselves to doubts and fears about the wisdom or even the existence of the providential plan, they

[115] BTW, I, 503.
[116] Bennett, 164.
[117] E. H. Stenning, "The Original Lands of Bishop Barrow's Trustees," Isle of Man Natural History and Antiquarian Society Proceedings, 5 (1942–1946), 122–145. See Barrow's sermon at his uncle's consecration, BTW, I, 500 et seq.

emphasize the stumbling block presented by the prosperity of the wicked. To them the main cause of religious skepticism was not, as we might have expected, an intellectual tradition at least as old as Montaigne, but rather the material circumstances of the doubter. Was it not obvious that in some cases the wicked, who are also very industrious, prosper?[118] Their prosperity induces "atheism or epicurism" among those who strive but are slightly less fortunate and provokes among the jealous the expression of comparable self-interest which as its main characteristic refuses to acknowledge the dictates of providence.[119] Arising from unbridled self-interest, often induced by Hobbes's principles which are unfortunately "so agreeable to flesh and blood," comes intellectual skepticism and finally atheism.[120] By atheism the latitudinarians almost invariably implied a life style spent solely in the pursuit of worldly rewards.

Who were these "crafty, ill-principled men," as Evelyn came to call them, who prospered at the expense of the virtuous? To Evelyn they were the great financiers like Sir Josiah Child or Sir John Banks who changed their religion or political principles as necessity and interest warranted.[121] In sermons where only general sentiment against the worldly could be expressed, the latitudinarians aim their barbed references at the courtiers, implying that religion would fare better if the gentry would pursue their business, or that the depravity of the mighty inhibits the growth of trade and national power.[122]

[118] *BTW*, III, 400; VII, 315–316; Wilkins, 86; *TW*, I, 375.

[119] *BTW*, VI, 442–443; IV, 114; V, 34.

[120] Moore, MS Dd. 14.15, f. 3ʳ, U.L.C. Preached at college chapel in 1671 and at Great St. Mary's, Cambridge. See Wilkins, 98–99, 291.

[121] See William Letwin, *Sir Josiah Child, Merchant Economist* (Boston, 1959), 23–24; D. C. Coleman, *Sir John Banks, Baronet and Businessman. A Study of Business, Politics and Society in Later Stuart England* (Oxford, 1963), 21, 67, 92, 146–147, and quotations from Evelyn's diary cited therein.

[122] Moore, MS Dd. 14.9, ff. 102–103, U.L.C.; *BTW*, III, 419–420; Moore, MS Dd. 14.15, f. 76, U.L.C.

Occasionally instability results from government run by men of money, such Barrow claimed were the causes of the Fronde; or disorder ensues "when the dregs get a chance to disrupt," an example used by Wilkins. As the supreme example of unbridled self-interest, Hezekiah Burton and Thomas Tenison nominated the French king, Louis XIV.[123] Latitudinarian discontent, when made explicit, was country-orientated in its opposition to the court and congenial toward independent entrepreneurs in its support for market forces and free trade as opposed to government- and magnate-controlled monopolies.

"The children of the world," as Tillotson called the wicked and the prosperous, were more cunning than the "children of light."[124] Their prosperity, seemingly gained at the expense of the church and its followers, goaded the latitudinarians into charging their congregations with the public-spirited pursuit of power and prosperity. Perhaps the evil beneficiaries of market forces and monopolies were figments of an envious imagination, but that is not important. What matters is that churchmen proposed their version of what was later to be called the Protestant ethic by Max Weber and others, wedded to the new natural philosophy, as an alternative to an already developed and operative ethic they loosely described as Hobbist. The spoils of business and commerce were already enjoyed by men who presumably knew how to compete in a ruthless and aggressive fashion. The latitudinarians were trying to stem a tide, not to hold back the growth of capitalistic forms of economic and social life, but to Christianize them. The spirit of a well-established capitalism had to be made compatible with the Protestant ethic, and this was the task of the latitudinarians. This point must be stressed even in the face of the seemingly convincing arguments to support a somewhat inverse relationship between Protestantism and capitalism put forward by R. H. Tawney and especially by Weber. In the period after 1660 low-churchmen leave us with

[123] Burton, 319; Tenison, *Sermon against Self-Love*, 25–27.
[124] *TW*, II, 489–490.

little doubt that they saw themselves embattled against the successful proponents of economic materialism and moral opportunism who already had developed a capitalist ethic.

In his famous analysis of 1922, Tawney used Tillotson's phrase, "the children of light," to describe, in sadness, the virtuous and socially conscious Christians of the seventeenth century who "denounced each new victory of economic enterprise as yet another stratagem of Mammon"[125] and who lost the campaign against capitalism out of ignorance. Tawney, of course, argued that Protestantism, particularly its Puritan variety, provided an essential ethic that enabled capitalism to be adopted "with whole-hearted enthusiasm."[126] Many English churchmen—and the discussion should no longer be waylaid by labels such as Puritan—saw themselves as the virtuous, the "children of light" whose fate Tawney bemoaned. Tawney failed to take account of what the sermon literature just discussed emphasizes constantly: the church posed its ethic of "sober self-love" in opposition to a fully developed possessive individualism already operative in society and best, but not exclusively, described, in the church's opinion, by Hobbes. We should not, therefore, join with Tawney, at least not on this occasion, or with Tillotson in mourning these particular "children of light"; they too were both cunning and resilient. As an alternative to prosperity gained through ruthlessness, the "children of light" refined natural religion and preached it in the city. Perhaps not all their congregations were convinced; but enough were. In latitudinarian natural religion they found a respectable and comforting Christian justification and explanation for activities in which they were already engaged and which could now be made compatible with the providential plan as revealed in nature.

The latitudinarians adapted Christianity to a market society by transforming it into a natural religion which would serve the needs of self-interest and make them compatible with the

[125] *Religion and the Rise of Capitalism* (New York, 1961, reprint ed.), 231.   [126] Ibid., 205.

dictates of providence. Science and natural philosophy were indispensable catalysts in that transformation. If nature could appear to operate according to certain mechanical principles directly controlled by a providential deity and discernible to man, then human desires for power and the acquisition of fortune could be allowed free expression. Nature would provide a model for Christianizing and harmonizing the operations of the "world politick," and the church, a comprehensive body of true Protestants, would apply that model and, in the workings of economic and social forces, reveal the operations of providence.

Churchmen possessed confidence that their plan for the "world politick" would work not simply because of their faith in the new science, but also because they assumed a sequence of historical events that were preordained and orchestrated by a sacred plan. History as recorded and predicted in Scripture foretold the eventual triumph of Protestantism and the institution by Christ of a thousand-year era of peace and godliness for the saints. There is evidence that many of the latitudinarians, quite probably including Newton himself, were millenarians. If we are surprised at their attitude it is because we have misunderstood the nature of seventeenth-century English millenarianism.[127] We have mistakenly assumed that such views were held only by radicals or fanatics, the understandable expression of their longing for justice and economic freedom. Anglican millenarianism can be found throughout the seventeenth century, and it served the interest of its advocates just as powerfully as did the millennial dreams of the radicals. To be sure, their paradises bore little resemblance to one

[127] See the remarks of William Lamont, "Richard Baxter, the Apocalypse and the Mad Major," *Past and Present*, No. 55 (1972), 72–74; and Jacob and Lockwood, 267. For a brief summary, Charles Webster, ed., *The Intellectual Revolution of the Seventeenth Century* (London, 1974), 8–12. For stress on the radical associations of millenarianism see Christopher Hill, "Sir Isaac Newton and His Society" in *Change and Continuity in 17th Century England* (London, 1974), 268–271.

another, and especially after the Restoration Anglican preachers appear reticent to discuss the approaching millennium in their sermons. Enough evidence is available from other sources to show that as late as the early eighteenth century millenarian speculations continued in church circles.

In a subsequent chapter I shall attempt to explain more precisely how millenarianism fitted into the latitudinarian conception of the "world politick" and the world natural. It is sufficient at this point to know that millenarian speculations lent credibility to the church's efforts to maintain its position and to reform society. The end of history could not be far away; indeed, Tenison explained, it is well known that men grow worse as judgment day approaches. In their sermons Tillotson and Moore warn that the time is drawing near when all will be set right and the wicked will be punished.[128] In the meantime men must curb their interest by industry and enjoy whatever prosperity God's providence allots to them. The church's task is to prepare men for their reward in the "new heaven and the new earth," and to do this the church must cultivate its interests and prosperity. Only a truly Protestant church, comprehensive and unchallenged by sectarianism, could prepare the saints for their promised reward in the millennial order. Throughout the Restoration the latitudinarians worked for comprehension and Anglican hegemony, always in opposition, they imagined, to the designs of antichrist, whom everyone knew to be Rome. In the reign of James II it appeared for a time that their efforts were doomed to failure.

[128] Tenison, *Sermon against Self-Love*, 14; and *A Friendly Debate* . . . (London, 1688), 18. *TW*, II, 575–576; III, 170–172, 247–248; Moore, MS Dd. 14.15, f. 75 et seq., U.L.C.

CHAPTER 2

The Church and the Revolution of 1688-1689

The English Revolution of mid-century provided the social context for the initial formulation of latitudinarian goals and social ideology. From their understanding of that experience the first generation of latitudinarians bequeathed to their successors certain fundamental principles, outlined in the previous chapter, which were to remain part of the liberal Protestant creed well into the eighteenth century. In 1688-1689, however, another Revolution threatened the church's interests and weakened its political and moral influence. Once again the experience of revolution shaped latitudinarian goals and fortunes.

The Revolution of 1688-1689, and the latitudinarian response to it, must be understood if we are to grasp the social meaning of Newtonianism. As a result of the Revolution, the latitudinarians became the ecclesiastical and intellectual leaders of the church, though they did not accept all that the Revolution Settlement proclaimed. During the events of 1688-1689, the attitudes of low-churchmen were ambiguous and uncertain. The reasons for their ambiguity are vital to any understanding of the subsequent social thought of the second generation of latitudinarian divines. Their newly acquired positions and influence made them public defenders of the Revolution and its Settlement. Yet doubts remained. The Revolution had seemed to the latitudinarians to affirm rapacious self-interest, coupled with indifference in religious matters. The task of Christianizing such a society, or preparing it for its historic mission, consumed latitudinarian energies after the Revolution.

What was needed, or so it seemed, were broad principles, based on the operation of nature, which would appeal to reasonable and sober men, regardless of their particular version of Protestantism. Once again the new mechanical philosophy proved relevant, only now the latitudinarians possessed Newton's version and application of that natural philosophy. The latitudinarians were forced, as a result of the Revolution, to rethink their social and political ideology and in turn to preach it to congregations whose fortunes and political interests had been altered and in most cases improved, also in consequence of the Revolution. The events that engulfed the church and English society during and after 1688–1689 in large measure explain why the latitudinarians preached a particular social and natural philosophy in the twenty or so years after the Revolution.

The church's interests were gravely threatened after James II's accession to the throne in 1685. His Catholicism, coupled with his attempt to undermine Anglican political hegemony, jeopardized the church's political power; it also exposed the inherent contradictions between militant Protestantism and the church's post-Restoration obsession with the theory of divine-right monarchy. After 1685 churchmen could not reconcile their insistence on divine right and passive obedience with their abhorrence of Rome, and increasingly they found themselves preaching against Catholic doctrine and practice while at the same time avowing their support for James II's legitimate political authority. This contradiction between Protestant pretensions and monarchical designs rendered the church's hierarchy politically ineffectual. With the notable exceptions of William Sancroft, Henry Compton,[1] and Gilbert Burnet, the church played the role of spectator in the planning

[1] David H. Hosford, "Bishop Compton and the Revolution of 1688," *Journal of Ecclesiastical History*, 23 (1972), 209–218.

and execution of the Revolution, and that event dealt harshly with the church's fortunes. The Revolution guaranteed the security of the church within the politico-religious establishment, but denied it the political power or social control it had possessed during the Restoration. The enactment of the Toleration Act, the establishment of Presbyterianism in Scotland, and the failure of the Comprehension Bill were merely obvious signs of the church's lessened importance. Beyond these signs, churchmen were convinced, lay a new social and intellectual order hostile to the church and to religion in general, charactized by dedication to the ruthless pursuit of political and economic interest. It was as if the Hobbists had secured their victory.

Like any sudden and rapid historical change, the Revolution and its outcome were unimagined in the early 1680s. As part of the vanguard of Tory reaction in that period churchmen,[2] even of latitudinarian mind, preached passive obedience with renewed vigor. Although unrelenting in their attacks on the "senseless Superstitions of Rome," churchmen like John Moore warned that "when we are so terrify'd about the events of things" we must resist under any circumstances recourse "to Cunning men for a Resolution." Their methods, like those of the exclusionists, are "downright sinful."[3] Moore and many of his colleagues were deeply troubled by the prospect of a Catholic monarch, yet they urged the passive acceptance of whatever might come.

Indeed deliverance came in 1688–1689, but by conspiracy and not passivity. Preaching before the new queen in 1690, Moore explained that providence operates in both the world natural and the "world politick" and determines our pros-

[2] See Robert Beddard, "The Commission for Ecclesiastical Promotion, 1681–84: An Instrument of Tory Reaction," *Historical Journal*, 10 (1967), 11–40.

[3] John Moore, "Sermon Preach'd before the Lord Mayor . . . Jan. 27, 1683/4," in Samuel Clarke, ed., *Sermons on Several Subjects* (London, 1715), 62.

perity even though sinful men "look upon the Prosperity of their Condition to be the effect alone of their Own Wisdom and good Management."[4] Between the dates of Moore's sermons on providence seven bishops had been imprisoned, widespread public disturbances had occurred, an invader had landed at Torbay on November 5, 1688, and a monarch had fled; his throne was then declared vacant and was filled by act of Parliament. The church's doctrines of passive obedience and divine-right monarchy had been rendered bankrupt. We can only imagine Moore's secret musings about the mysterious ways of God's providence.

Although unfortunately we possess no record of John Moore's private thoughts, we do have some revealing information about those of his latitudinarian colleagues. This information allows us to establish with some certainty the reaction to the Revolution of prominent Anglicans such as John Evelyn and Thomas Tenison, who were either directly or indirectly associated with the Boyle lectureship. The Boyle lectures, established by the will of Robert Boyle (d.1691), became one of the important means low-churchmen used to communicate their natural religion based on Newtonian principles and to demonstrate its relevance to the issues that confronted the church after the Revolution.

Our evidence establishes, it would appear beyond reasonable doubt, the low church's profound ambivalence toward the Revolution as it was occurring. Only after the settlement rendered the Revolution a fait accompli did latitudinarian churchmen accept the new order and seek to explain it by reference to the providential plan. Their ambivalence carried over, however, and manifested itself in distrust and uncertainty about the political and economic order sanctioned by those events. The emergence of a new, less rigid, economic structure[5] presented unforeseen opportunities for the expression of

[4] Ibid., 120–121.
[5] Christopher Hill, *The Century of Revolution, 1603–1714* (Edinburgh, 1963), 263–266.

self-interest. Latitudinarian social theories developed during the Restoration assumed greater relevance after 1688–1689 but not necessarily greater acceptability to the self-interested.

This is not to say, however, that during the reign of James II, his policies ever won support among low-churchmen. He made two irredeemable mistakes. He attempted to undermine the Protestant religion by placing Catholics in positions of power in the government, military, universities, and sundry ecclesiastical offices, and he extended his power far beyond the court into the towns and boroughs—the centers of power previously reserved for the gentry.[6] His policies, such as the Declaration of Indulgence and the bypassing of the Test and Corporation Acts constituted a bid for absolute power that fundamentally threatened the position of the church, the Protestant aristocracy, and the gentry. Common to the handful of noblemen[7] and to the important representatives of the gentry who negotiated with William for the preservation of English rights and religion was the belief that unless James were stopped, all order would be lost and the political, economic, and religious power they had worked so hard to preserve after 1660 would be permanently undermined.[8]

The beneficiaries of the Revolution were not always the same people who had most fiercely resented the encroachment of King James. After 1688 the interests of the country gentry, in particular the less prosperous members of that group, often gave way, as J. H. Plumb states it, to "the authority of certain men of property, particularly those of high social standing either aristocrats or linked with aristocracy, whose tap root was in land but whose side roots reached out to commerce, industry

[6] J. H. Plumb, *The Growth of Political Stability in England, 1675–1725* (London, 1967), 58, 62.

[7] J. P. Kenyon, *The Nobility in the Revolution of 1688* (Hull, Yorkshire, 1963), 11.

[8] Lucille Pinkham, *William III and the Revolution of 1688* (Cambridge, 1954), esp. 239.

and finance."⁹ The lesser gentry were not the only ones whose aspirations were thwarted in 1688. For the so-called "mob" or "rabble," 1688 changed little. Considerable evidence exists to indicate that, much to the apprehension of their superiors, the crowd, particularly its artisan segment, engaged in meaningful political activities from mid-November 1688 to February 1689.¹⁰ Crowd disturbance even in support of William, however, only confirmed the fears of the London magistrates and prompted Parliament to give governing power to William. It is difficult to know the precise motives behind this crowd activity. Certainly the conditions of the poor throughout the Restoration period were precarious; charity was the preserve of the church and was often given only as an award for apparent devotion to it.¹¹ Despite the discontent of the poor, their situation remained unaltered by the Revolution.

Within the church, 1688–1689 brought to the surface fundamental disagreements among its various factions. Although, as we shall see, many latitudinarians questioned the wisdom of events, nonetheless John Tillotson, Edward Stillingfleet, William Lloyd, John Moore, Simon Patrick, and Edward Fowler, to name only the most prominent, all accepted high ecclesiastical offices. They did so in many cases at the expense of other churchmen who, apparently as a matter of conscience, refused to take the oath and lost their positions. Led by Archbishop Sancroft, approximately four hundred clerics refused to take the oath to William and Mary, and the church fell into schism. Many others, despite their sympathy for Sancroft's

⁹ Plumb, 69. Cf. J. R. Jones, *The Revolution of 1688 in England* (New York, 1972).
¹⁰ W. L. Sachse, "The Mob and the Revolution of 1688," *Journal of British Studies*, 4 (1964), 36. See also Stowe 370, B.L., for constant reference to fear of "rabble" in the journal of Thomas Bruce, second earl of Ailesbury.
¹¹ H. W. Stephenson, "Thomas Firmin, 1632–94," 3 vols. (Ph.D. diss., Oxford University, n.d.), 130–131.

position, remained within the church. Francis Atterbury, William Jane, Henry Aldrich, Bishops Sprat, Compton, and Sir Jonathan Trelawny, all high-churchmen by the late 1690s if not before, pressed their case at Convocation and provoked a crisis that publicly split the church throughout the twenty-five-year period after the Revolution.[12] A churchman's attitude toward the Revolution, the meaning of the Toleration Act, and the Protestant succession would shape his politics up to 1714. In Anne's reign church politics became nearly as heated as parliamentary politics; indeed the two cannot be separated. By 1710 low-churchmen avoided even social events where their high-church brethren might be present.[13]

The split between high church and low church entailed more than politics and political theory; it extended to intellectual life as well. After the Revolution high-churchmen increasingly confined their sermons to assaults on the new order, which in some instances devolved into a carping and obscurantist opposition to everything new and modern. In their opposition to the post-Revolution order they often failed to offer intellectual leadership to the church, and increasingly confined themselves to supporting ancient learning and attacking the latitudinarian proponents of the new, modern learning. The backers of Charles Boyle in the controversy with the Newtonian, Richard Bentley, over Boyle's translation of *Epistles to Phalaris* (1694) were prominent high-churchmen, Atterbury and Aldrich among others,[14] and in the ensuing academic squabble over authenticity of the *Epistles,* political positions merged with scholarly and personal antagonisms. Briefly stated, Bentley claimed that the *Epistles* was a forgery, and Boyle, along with other supporters of the ancients, counterattacked and accused

[12] George Every, *The High-Church Party, 1688–1718* (London, 1956).

[13] Geoffrey Holmes, *British Politics in the Age of Anne* (London, 1967), 21, passim.

[14] See W. G. Hiscock, *Henry Aldrich of Christ Church, 1648–1710* (Oxford, 1960); W. R. Ward, *Georgian Oxford, University Politics in the Eighteenth Century* (Oxford, 1958). Cf. G. V. Bennett, *The Tory Crisis in Church and State 1688–1730* (Oxford, 1975), 38–43.

Bentley of declaring "open war against Phalaris, and all his Party."[15]

Phalaris was the tyrant of Agrigentum who was finally exiled for his cruelty. I would imagine that in the controversy over the authenticity of his letters, Phalaris became a symbol; for the supporters of authenticity, of their sympathy for the exiled king, James II, and for the detractors of the *Epistles* and of the superiority of ancient learning, of their acceptance of the new, post-Revolution order.

In its profound discontent, therefore, the high-church party managed to ally opposition to the new order with support for ancient learning. By comparison, the latitudinarians appeared to be aggressively modern. Of all high-churchmen perhaps only Jonathan Swift possessed the talent to transform the sources of high-church discontent into telling and poignant criticism of his society and to offer, in the process, alternative ideals for social and moral behavior. The high-church faction failed to use the new learning, and therefore, the discoveries of Newton, because their intellectual life was conditioned by opposition rather than by the desire to address themselves to the new social and political values sanctioned by the Revolution.

As the cornerstone of their political theory the high-church faction, led by Atterbury, argued that the church is independent of the state, that convocation possesses in the ecclesiastical sphere the same power possessed by Parliament in the political and economic, and finally that monarchy must be independent and truly sovereign. Of course, the English church had never been independent of the state, but after the Revolution it was less so than ever before in its history. By the reign of Anne ecclesiastical preferment depended almost entirely on the secular political forces then in power.[16]

[15] Charles Boyle, *Dr. Bentley's Dissertations on the Epistles of Phalaris and the Fables of Aesop, Examined* (London, 1698), 30.
[16] See G. V. Bennett, "Robert Harley, the Godolphin Ministry, and the Bishoprics Crisis of 1707," *English Historical Review*, 82 (1967),

Given their fears after the Revolution, it is hardly surprising to find churchmen in 1688–1689 deeply disturbed and distrustful about the course taken by events from the summer of 1688 onward. To the conspirators who formed the hard core of William's supporters the moral issue presented by his invasion was easily solved. They relied on contract theory to argue that James, by his politics and then his flight, broke the contract between king and people and thereby nullified his right to the crown. But for the majority of churchmen the dilemma presented by James's policies could not be solved by reference to contract theories.[17] Committed to their Restoration teachings of passive obedience and divine-right monarchy, the church found insoluble the moral problems presented by the king's open subversion of the Protestant establishment.[18] When churchmen accepted the Revolution they did so not as much out of conviction as of necessity. In this process of acceptance they weighed up the benefits and deficits presented by the new order and searched for theories to justify the constitutional upheaval and in turn to apply to the kind of society that they believed came into being as a result of the Revolution.

The latitudinarian acceptance of the Revolution warrants close examination. Writing about 1688–1689 in his *History*, Bishop Burnet observed that "the doctrines of passive obedience and non-resistance had been carried so far, and preached so much, that clergymen either could not all on the sudden get out of that entanglement, into which they had by long thinking and speaking all one way involved themselves, or they were ashamed to make so quick a turn."[19] To Burnet, who as early as 1685 had gone to the Continent presumably

726–746; and G. V. Bennett, "Conflict in the Church," in Geoffrey Holmes, ed., *Britain after the Glorious Revolution, 1689–1714* (London, 1969), 155–175.

[17] G. Straka, *The Anglican Reaction to the Revolution of 1688* (Madison, Wis., 1962), viii.

[18] N. Sykes, *Church and State in England in the XVIIIth Century* (Cambridge, 1934), 28–29.

[19] G. Burnet, *History of His Own Time* (London, 1839), 500.

with the solution to this dilemma already formulated, the effect of the Revolution on religious thinkers could be summed up in one sentence. But to John Evelyn, Thomas Tenison, William Lloyd, Daniel Finch, second earl of Nottingham, his brother Heneage, and many others, the occurrence of a revolution brought to the surface their deepest fears about the structure of the body politic, forced them to come to terms with a changed order, and led them to formulate theories that would explain political upheavals and provide some security against their recurrence.

For an example of their dilemma we turn to the diary of John Evelyn. He vividly reveals his reactions to events from 1687 to 1690 and thereby provides considerable insight into the day-to-day reactions of that moderate section of the church with which he was affiliated. Evelyn's religious sensibility had been formed in the 1650s and 1660s. Although not an aristocrat himself, he strongly identified his economic and social aspirations with those of the aristocracy and accepted as moral values the aristocratic notions of virtue and honor. During the Commonwealth period, like so many aristocrats or royalists, he retired for a time to France, and later returned to England only to remain in virtual seclusion. Symptomatic of this withdrawal is Evelyn's own writing. He was a major contributor to the "garden literature" of the period—writings that satisfied the needs of men who had been forced to tend to their gardens while others looked after political and social affairs.[20]

Evelyn became an avid propagandist for the Restoration and seems to have prospered under the returned monarchy. Yet Evelyn was never simply a sycophant of the Stuart court. His interests extended to the scientific endeavors of the day, and his social life often centered around the *virtuosi*. He became a friend of Robert Boyle's, and Boyle in his will desig-

[20] For an interesting discussion of this phenomenon see Maren-Sofie Røstvig, *The Happy Man: Studies in the Metamorphoses of a Classical Ideal, 1600–1700*, I (Oslo, 1954), 56–66.

nated him, along with Thomas Tenison, a trustee of the Boyle lectures. During the 1680s Evelyn's closest social links were with the moderates of the church, with *virtuosi* such as Boyle and Samuel Pepys and, of course, with aristocratic families such as the Sunderlands, the Finches, and the Hydes.

Although Evelyn had gained preferment under James and thereby revitalized his weakened financial state,[21] his diary indicates that he had no sympathy with James's policies toward either Catholics or Dissenters. Evelyn was a firm supporter of the Anglican establishment and constantly noted the sermons by Tenison, Stillingfleet, Sharp, Patrick, and others, warning Protestants of the danger of losing their faith and cataloguing the evils of popery.[22] In 1686, Evelyn wrote that England had "fallen from its antient zeal and Integritie," and that unless the country were restored to its ancient mission, its destruction was imminent. He viewed the recent persecution and dispersion of the French Huguenots as a portent of this impending fate.[23]

In this melancholy state of mind, every Sunday Evelyn attended St. Martin's-in-the-Fields, where he heard his friend Thomas Tenison, pastor of the parish, defend Protestant teaching on Scripture and against transubstantiation, as well as urge his congregation to put their trust in God's providence.[24] His power would be the only means to effect salvation in the face of the antichrist.

Another danger concerned Evelyn and Tenison.[25] It was the "vanity of riches"—the materialism of "this vicious age" that Evelyn believed contributed to the confusion and irreligion surrounding him.[26] Yet did not Evelyn, and even more

[21] W. G. Hiscock, *John Evelyn and His Family Circle* (London, 1955), 143–144.
[22] E. deBeer, ed., *The Diary of John Evelyn* (Oxford, 1955), IV, 500–504, 526–527.
[23] Ibid., 512, 508–525.
[24] Ibid., 531, 537, 539.
[25] Ibid., 564.
[26] Ibid., 584–585, May 23, 1688.

so the aristocratic circle within which he moved, accumulate wealth and continue to do so after 1688? To Evelyn and to the church, prosperity did not conflict with religion, provided the beneficiary acknowledged his debt to God's providence. It was the *vanity* of riches that perturbed churchmen, the refusal of the moneyed and landed interests to give even a passing nod to God's beneficence and in return for his generosity to acknowledge the role of the church as the preserver of that stability without which prosperity became a tenuous business at best. Week after week Evelyn observed this vanity among his fellow worshippers in the wealthy parish of St. Martin's.

By 1688 the fears that Evelyn had expressed earlier were being confirmed. On May 25 he met with Tenison, Pepys, Boyle, and several friends to discuss the imprisonment of seven bishops. On the same day Evelyn noted news of a terrible earthquake that had destroyed Lima, Peru.[27] These two events, apparently so disparate, were intimately connected in Evelyn's mind. The earthquake could be a natural sign from God, foretelling an impending political disaster. On July 12, Evelyn recorded that members of the Presbyterian and Independent "party" had joined the Privy Council, and then added the ominous assessment: "To effect their owne ends which was not evidently the utter extirpation of the Church of England: first, and then the rest would inevitably follow." To Evelyn the church was the guarantor of stability within society. If its power was broken then the very structure of society would crumble.

Evelyn and his contemporaries based their fear upon historical fact. For them the Commonwealth remained a haunting memory. Even the youngest of Evelyn's associates knew intimately the dislocation produced in their world by the civil wars and Interregnum. In the summer of 1688, Evelyn mourned the "calamities that befell the Israelites upon breach of the Covenant, God made with them." As the political

[27] Ibid., 585–586.

situation deteriorated, Evelyn's fears worsened. Even though Tenison informed him on August 10 of the Prince of Orange's intended coming,[28] Evelyn noted on September 22: "Earthquakes had now utterly demolished the ancient *Smyrna* and several other places, both in Greece, Italy, and even the Spanish Indies, forerunners of greater Calamities: God Almight[y] preserve his Church, & all who put themselves under the shadow of his Wings, 'til these things be overpast."[29]

A diary entry made two weeks later clarifies what Evelyn meant by this plea. The people have been made so desperate by the king's policies that they now desire "the landing of that Prince, whom they looked on as their deliverer from popish Tyrannie."[30] From this time, Evelyn's diary entries reveal the attitudes and fears of a spectator forced to watch a disturbing, unfortunate, yet inevitable drama.

Shortly before William landed, Evelyn's only reaction was "the apprehension . . . that his Majesties Forces would neither at land or sea oppose them with the vigour requisite to repell Invaders."[31] When William had in fact landed, Evelyn was neither his supporter nor his opponent. Like the vast majority of the church's clergy, he stood on the sidelines, fearing that the instability produced by the invasion would undermine the very structure of society.

In a letter to the countess of Sunderland dated December 22, 1688, Evelyn partially reveals the kind of social relations he believes to be essential for the welfare of the nation:

And it is an eternal truth, and can never be otherwise, that true honour and happiness and the things which we seek (would consummate our felicity and bound our further pursuits), is not to be found in the things which pass away like a dream when we

[28] Ibid., 592; Tenison told Simon Patrick on August 7; *The Autobiography of Simon Patrick, Bishop of Ely* (Oxford, 1839), 137.

[29] deBeer, ed., IV, 598.

[30] Ibid., 600.

[31] Ibid., 600, Oct. 6, 1688.

awake; but in a brave and generous soul, that having those advantages by birth or laudable acquisition, can cultivate them to the production of things beneficial to mankind, the government, and eminent station in which God has placed him.[32]

Writing to the sympathetic countess, Evelyn acknowledges his belief that only the aristrocrat, made not necessarily by birth but also by "laudable acquisition" and possessing a vision that transcends the "things which pass away like a dream," can bring about progress in society and government. He tells her that never was this principle so true as in the "approaching revolution." It appears that Evelyn saw in the events of November and December 1688 the possibility of a major reorganization of society producing a world where the dreams of the present would distract men from their truly religious goal of bringing progress to mankind. Concomitant with this new organization would be the appearance of "lesser men" who would fashion the new mundane order. Evelyn appears to fear the new moneyed interests—those very interests that were to come into their own after 1688.[33] He fears them not because of their prosperity, but because, to Evelyn, they had lost sight of the truly religious motivation behind human endeavor—an essential motivation if the Reformation of religion and society was to be fulfilled.

It is hardly surprising then that in April 1689, when the Revolution had been secured, Evelyn judged it as having been "managed by some crafty, ill principled men: The new Pr[ivy] Council having a Republican Spirit, and manifestly undermining all future Succession of the Crown, and Prosperity of the Church of England."[34] At the time Evelyn made this condemnation the Convention Parliament was dominated by the Whigs.

By December, however, the political situation had changed

---

[32] John Evelyn, *The Diary and Correspondence* (Bohn ed., London, 1859), III, 292–293.
[33] Hill, 266–271.
[34] deBeer, ed., IV, 635.

noticeably. Evelyn allowed himself a glimmer of hope, noting that the Old Testament prophecies show what wonderful things are to come if we live as Christians.[35] Evelyn's returning optimism was quite probably due to the recently restored power of the church. In August 1689, William had decided to form an alliance with the party "between the 2 extremes."[36] He had shifted his policy because he feared the extremes to which the Whigs might go. He was shrewd enough to know that men who had deposed one king might not scruple at deposing a second. In order to pursue the Continental war and to ensure his power at home, William was determined to be a strong monarch and to ally only with the political faction that benefited those aims.

In December 1689, Evelyn could once more plan for the "wonderful things" that were in store for Christians as long as the church could retain its political and social influence. For the mission of the church in his time had been to bring order to society: religion was the social cement.[37]

If men recognized God's providence and government in the "world politick" then indeed the millennium, as Evelyn understood it, would not be far off. But if the events of 1688 and 1689 were repeated or if the secular and worldly men who deposed kings and crowned new ones were allowed the upper hand, then the destruction that Evelyn had so desperately feared would become a reality. The delicate balance in Evelyn's mind between the hope for a new order in society, one that would allow for the accomplishment of the Scriptural prophecies and harmony in human affairs, and the fear that the delicate fabric of order could be torn asunder by crass, evil, and "atheisticall" men, is best expressed in Evelyn's own words to his son:

[35] Ibid., 653, Dec. 8, 1689.
[36] Quoted by G. V. Bennett, "William III and the Episcopate," in G. V. Bennett and J. D. Walsh, eds., *Essays in Modern English Church History* (London, 1966), 117.
[37] See the sermons of William Fleetwood in *A Compleat Collection* . . . (London, 1737).

But as I have often told you, I look for no mighty improvement of mankind in this declining age and catalysis. A Parliament (legally called) of brave and worthy patriots, not influenced by faction, nor terrified by power, or corrupted by self interest, would produce a kind of new creation amongst us. But it will grow old, and dissolve to chaos again, unless the same stupendous Providence which had put this opportunity into men's hands to make us happy, dispose them to do just and righteous things, and to use their empire with moderation, justice, piety, and for the public good. . . . These [difficulties relative to preferment] and sundry other difficulties will render things both uneasy and uncertain. Only I think Popery to be universally declining, and you know I am one of those who despise not prophesying; nor whilst I behold what is daily wrought in the world, believe miracles to be ceased.[38]

The fulfillment of the prophecies and the working of miracles in the civil polity depended entirely on the power the church could exercise in steering men along the paths of righteousness.

The precarious position of the church, indicating to churchmen the precarious nature of the new political order, also disturbed Thomas Tenison. He, far more than Evelyn, had taken an active role in opposing the policies of James II. In 1688 he edited *Popery not founded on Scripture*, a collection of essays by prominent churchmen attacking almost every aspect of Catholic doctrine and practice.[39] In another tract on the "Protestant method" Tenison firmly placed himself in the theological tradition of moderation first formulated in Anglican circles by William Chillingworth. Tenison argued that Scripture alone was the rule of faith and not the authority of any church.[40]

[38] Evelyn, *Correspondence*, 290, Dec. 18, 1688.
[39] E. Carpenter, *Thomas Tenison* (London, 1948), 67. On page 68, Tenison is listed as the translator of J. de la Placette's *Of the Incurable Scepticism of the Church of Rome*. Henry Wharton in his diary (MS 1169*, Lambeth Palace Library), notes that he translated it for Tenison.
[40] Tenison, *The Difference Betwixt the Protestant and Socinian*

Despite Tenison's firm stand in the 1680s against measures he felt violated the welfare of the church, when the Revolution came its acceptance presented certain real difficulties. Unfortunately, we do not possess any diary by Tenison.[41] In trying to reconstruct his attitudes toward events it is necessary to rely upon Evelyn's rather detailed account of his sermons and on the content of the Scriptural text used in each sermon.

In 1686, Tenison became increasingly worried about the growing influence of Catholics in public affairs. Preaching against this situation, he urged, as had John Moore and many other of his colleagues, that men put their faith in God's providence.[42] In January 1688, Tenison argued for "the good and necessity of Order and government."[43] Yet in May, Tenison added his signature to the bishop's petition against promulgating the Declaration of Indulgence and aided activities aimed at stopping its circulation. But at the end of October, when rumors of William's coming were rife, he preached on the theme that "falsehood and not truth has grown strong in the land" (Jer. 9:3). In this sermon Tenison warned that men "proceed from evil to evil" and that they have lost sight of God.[44] To Tenison, apparently opposition was one thing but open rebellion quite another. On December 2, eight days before James's flight, we hear that once more Tenison appealed to God's providence to save men.[45] The text (Ps. 36:5, 6, 7) tells us that the judgments of God "are like the great deep."

Methods . . . (London, 1687), esp. 58. For a good discussion of Chillingworth's theology, see Robert R. Orr, Reason and Authority: The Thought of William Chillingworth (Oxford, 1967).

[41] A Tenison commonplace book in the Forster Collection of the Library of the Victoria and Albert Museum reveals his interest in science.

[42] deBeer, ed., IV, 531, Dec. 12, 1686.

[43] Ibid., 566.

[44] Ibid., 601-602.

[45] Ibid., 608.

Tenison now appears to be pondering the mysterious ways of God.

James II fled London on December 10. Two days later, Evelyn tells us that Tenison preached on Isaiah 8:11.[46] The text, "For the Lord spoke thus to me with his strong hand upon me, and warned me not to walk in the way of his people," is not particularly revealing unless it is given in its entirety, saying: "Do not call conspiracy all that this people call conspiracy, and do not fear that they fear, nor be in dread" (Is. 8:12). It appears that Tenison has grudgingly given his assent to the Revolution. Faced with a situation in which the now kingless country had no choice but to accept what often appeared to Tenison and many others as a conspiracy, he urged compliance.

At this stage his own compliance was not wholehearted. On the crucial constitutional and legal question whether or not James had abdicated Tenison appears to have expressed his feelings in a sermon delivered the day before the House of Commons rendered its decision that the king had in fact abdicated. The impact of this decision that crushed any hopes of reconciliation with James, whom many still considered to be the rightful, although absent, monarch, is presaged in Tenison's sermon "shewing the universal Corruption of men's hearts, and exhorting to a serious watchfullnesse over them."[47] In the early months of 1689, Tenison returned to this theme, referring constantly to man's wickedness. On February 17 he urged the "necessity of begging pardon of God, to clense us from our seacret sinns," and on March 2 he asked, "For what will it profit a man if he gains the whole world and forfeits his soul?" (Matt. 16:26).[48] In contrast to Tenison's ob-

---

[46] Ibid., 610; deBeer thinks that the reference is doubtful. But if Isaiah 8:11 is seen in the light of Isaiah 8:12, the text becomes immediately relevant.

[47] Ibid., 616, Jan. 27, 1688/9.

[48] Ibid., 623, 627.

servations stand those of Gilbert Burnet, perhaps William's firmest church supporter, who reminded his congregations of the necessity "to walke worthy of God's particular and signal deliverances of this Nation and Church."[49]

As in the cases of Evelyn and Tenison there were other loyal supporters of the church who reacted with ambivalence to what Burnet saw as "God's particular and signal deliverances." The behavior of Daniel Finch, second earl of Nottingham and one of the church's most loyal supporters in this period, reveals a similarly ambivalent pattern. His opposition to James's policies of Catholic supremacy was well known.[50] Yet, in early 1688 when he was asked to lend support to William's plans for invasion, he refused. His opposition was based upon legal and moral conviction.[51] Nottingham's allegiance arose from his understanding of God's laws for the conduct of subject toward monarch. Because of his view of God's power in human affairs, the prospect of a revolution posed for Nottingham a pressing moral problem.

In an effort to find a solution he turned to two important and moderate churchmen, Edward Stillingfleet and William Lloyd. He asked them if one could "in Conscience endeavour to oppose by Force a manifest designe of destroying our religious and civil rights and liberties, though such an attempt by Force to defend them could not be justifyd by the known standing Laws of the realm?"[52] Significantly, they answered his question negatively; Nottingham played no part in the Revolution.[53]

When the Prince of Orange was entrenched in the country and James had fled, Parliament assembled to debate the ques-

[49] Ibid., 623, Feb. 21, 1688/9.
[50] H. G. Horwitz, *Revolution Politicks: The Career of Daniel Finch, Second Earl of Nottingham, 1647–1730* (Cambridge, 1968), chap. 4.
[51] Ibid., 52.
[52] Ibid., 53.
[53] Ibid., 143.

tion of William's right to the kingship. The ensuing debate, held at a joint session of the Houses, vividly revealed the conflicting political and ideological positions on the question of hereditary succession.[54] The earl of Clarendon, who later refused the oath to William and became one of the leaders of the extreme high-church position, refused to accept even the notion of a contract between king and people. He argued that "this breaking of the originale Contract is a language, that had not been long used in this place, nor known in any of our Law Books, or publick Records." The monarch is king before he is crowned and, afterward, "there is a naturall Allegiance due to him from the Subject immediately upon the descent of the Crown upon him."[55] Clarendon's argument in support of hereditary succession rested upon one basic notion, that of an indestructible "natural allegiance." To Clarendon and his followers the denial of hereditary succession would make the throne elective and thereby turn the government into a Commonwealth. He argues that "election is not by God's approbation."[56] A commonwealth could therefore exist only in direct disobedience to God's plan.

Nottingham and his brother, Heneage Finch, who represented Oxford University at the Convention Parliament, also opposed the Parliament's desire to depose James.[57] Their scruples were not as great as Clarendon's, however, and in the end they conformed to the new regime. The Finches were prominent in church affairs after the Revolution, and they aided the promotion of such promising young latitudinarians

---

[54] W. Cobbett, ed., *Parliamentary History of England . . . to the Year 1803*, 36 vols. (London, 1806–1820), V, 66–107, and Egerton 3362, B.L. A convenient reprint is *The Debate at Large between the House of Lords and House of Commons 1688* (Dublin, 1972).

[55] Egerton 3362, f. 38–38v, B.L.

[56] Egerton 3362, ff. 72v, 91, B.L.

[57] MS Rawl. A. 77 (24), Bodleian; Stowe MS 364, f. 62 et seq., B.L. Both probably favored William's appointment as regent, Horwitz, 74 and note.

as William Wotton and probably Richard Bentley.[58] Heneage Finch also took an active interest in the Boyle lectures.[59]

For a time this church faction in the Convention Parliament advocated making William regent and thereby preserving the hereditary succession. But he would have none of it. In the end the arguments of the Whigs, based on necessity and contract theory, prevailed and the Revolution was secured.

Churchmen continued to have misgivings about its wisdom and its legality. Humphrey Prideaux, dean of Norwich in the 1690s and a peripheral member of the moderate faction, was convinced that the Toleration Act "hath almost undone us. . . . Phanaticisme hath got the prevalincy in corporations, and the gentlemen must humour them this way or else they will not be chosen."[60] White Kennett, one of the most avid low-church propagandists and trustee of the Boyle lectureship during the reign of Anne, began his career after the Revolution as a high-churchman.[61] He only changed sides when he realized the fanatical extremes to which high-church opinion went on the Convocation question.

Many of the low-churchmen who led the church in the reign of William and controlled it to a lesser extent in the reign of Anne also harbored a certain scrupulosity over the constitutional settlement. In early 1690, Lloyd, Tenison, Francis Turner, and George Hickes, dean of Worcester, met at Clarendon's home for dinner. In his diary Clarendon recorded their revealing conversation: "After dinner . . . we fell upon the subject of the times, and concerning the Bishops who were to be deprived. Dr. Tenison owned, there had been irregularities in our settlement; that it was wished things had

[58] W. Finch to John, bishop of Norwich, Sept. 4, 1692, Tanner MSS, XXV, f. 399, Bodleian.

[59] See Chapter 4.

[60] *Letters of Humphrey Prideaux . . . to John Ellis, Sometime Under-Secretary of State, 1674–1722* (London, 1875), 154, June 27, 1692.

[61] G. V. Bennett, *White Kennett, 1660–1728, Bishop of Peterborough* (London, 1957), 12–13.

been otherwise, but we were now to make the best of it, and to join in the support of this government, as it was, for fear of worse." Lloyd then admitted that originally he had favored a regency, but now that William was king, "he [Lloyd] looked upon acquisition to beget a right."[62]

Ambiguity toward the Revolution and its Settlement made churchmen suspicious and even hostile toward those who prospered under the new regime. Despite their own sudden prosperity, the latitudinarians were caught in a dilemma. It was their task, as political leaders, to preserve the *via media*, to create a delicate balance among the many conflicting political forces that had made the Revolution or had repudiated it. The lower clergy, many of whom held high-church sympathies, had to be placated. Republicanism still remained a threat, as we shall see in our discussion of the freethinkers. More immediately the church recognized and feared the challenge posed by the growing power of the Dissenting sects.

Yet how would the church address itself to its political and social tasks if the Revolution had been of doubtful morality? During the Restoration the latitudinarians had preached to the vain and prosperous and presented them with a social philosophy that offered both worldly success and salvation. They had been able to do so because they believed that providence made all operate in conformity with a preordained plan, revealed both in the natural order and in history. If the latitudinarians were to engage in the urgent task of Christianizing a new order that to their minds was dangerously secular it was incumbent upon them to find the hidden meaning of the Revolution. They were able to do so because their social philosophy was sufficiently flexible and broad that it could accommodate political changes, even ones which threatened the church's interests.

[62] Samuel W. Singer, ed., *The Correspondence of Henry Hyde, Earl of Clarendon and of His Brother, Lawrence Hyde, Earl of Rochester; with the Diary of Lord Clarendon, from 1687, to 1690* (London, 1828), II, 300.

By 1691 the moderates controlled the church. Tillotson, who eventually replaced Sancroft as archbishop, Stillingfleet, Patrick, Lloyd, Tenison, and Gilbert Burnet exercised increasing control within the church, and in 1695 all church preferment fell into the domain of the latitudinarians and remained theirs until the reign of Anne.[63] Their rise to power had been as much a matter of luck as of skill, and the task now fell on them to explain the church's position in the new order, to devise new arguments in support of monarchy, and to promote their natural religion with renewed vigor.

What were they to make of events of such questionable rectitude that brought with them preferments and prosperity? Surely there must be a purpose or logic not apprehended at the time. On the existence of a divine plan in history rested the entire fabric of latitudinarian natural religion—the work ethic, enlightened self-interest, obedience to church and state, royal supremacy, the evil of rebellion and disobedience, and finally the fulfillment of the Reformation. Just as the providence of God had played a crucial role in latitudinarian thinking during the Restoration, the same providence provided a necessary and suitable explanation for the Revolution.

Churchmen avoided any justification of the Revolution that rested on contract theory. To their minds the theories of Locke and the other contractualists disregarded the necessity of God's active participation in the affairs of men. All de facto theories were absolutely unacceptable to churchmen, quite possibly because they associated any such theories with what they would have called Hobbism.[64] As Gerald Straka has shown, the arguments of the church in support of the Revolution rested entirely on the concept of a providential God. Providential right displaced divine right as the raison d'être for monarchy; indeed the entire Revolution became a

[63] Bennett, "William," 102–124.

[64] Cf. Quentin Skinner, "Conquest and Consent: Thomas Hobbes and the Engagement Controversy," in G. E. Aylmer, ed., *The Interregnum: The Quest for Settlement 1646–1660* (London, 1972), 97–98.

manifestation of God's plan in the universe.[65] The Restoration notions of passive obedience and the divine right of hereditary succession now became anachronistic.

Although there were many loyal Anglicans in Parliament such as Isaac Newton, who urged their constituents to take the Oath of Allegiance because the doctrine of passive obedience obliged them to do so,[66] this argument had little weight after 1689. The need for a complete rethinking of the church's teachings was painfully obvious. The moderate faction turned to Scripture and from it drew illustrations to prove that in the Revolution, God had fulfilled his word. William was seen as the new David, the French monarch became the antichrist, and the Continental war became a holy war. The providence of God gave support to the church's political theory and, as we shall see, this notion became absolutely essential to latitudinarian natural philosophy. With this notion the church explained and justified the Revolution and set the foundation for later arguments, developed in the Boyle lectures and designed to promote social and political order and stability.

The very fact of revolution could not be accepted unless it could be shown that even during social and political upheaval the will of God operated in the affairs of men. Even if churchmen who watched the events of 1688–1689 could "see not [at the time] a reason of the Alterations that were made," yet any believing Christian knew that God works in mysterious ways. The necessity of restoring order out of this confusion induced churchmen to stress God's role in the universe as an all-wise governor, a providential deity who governed with supreme order and logic.[67] This formulation of God's relationship to the universe had been an important part of Restoration philos-

[65] G. Straka, "The Final Phase of Divine Right Theory in England, 1688–1702," *English Historical Review*, 77 (1962), 638–658; and his *Anglican Reaction to the Revolution of 1688* (Madison, Wis., 1962).

[66] H. W. Turnbull, ed., *The Correspondence of Isaac Newton*, III (Cambridge, 1961), 12–13; cf. Millicent B. Rex, *University Representation in England, 1604–1690* (London, 1954).

[67] Straka, *Anglican Reaction*, 43.

ophy and theology as developed by natural philosophers such
as Robert Boyle and by the early latitudinarians; the post-1688
latitudinarians gave this notion widespread exposition and acceptance. The God of order, governor of the universe, became the cornerstone of the church's political and social teaching. God's providence operated in every aspect of reality—in
the natural order and in the world of human affairs men observed the preserving providence of God operating the laws
of nature and directing human affairs.[68] If a monarch violates
the laws of God or of the state, churchmen argued that the
subject had the right to do whatever was necessary to set
human affairs back on the course most closely conforming to
God's providential design.[69] In those circumstances providence
even sanctioned conquest and rebellion. William Lloyd found
such means acceptable even if the conqueror "means nothing
perhaps, but the satisfying of his own Lust. But though he
knoweth it not, he is sent in God's Message."[70] In the arguments developed by churchmen, William became, albeit unwittingly, the servant of God's plan for the English nation.

Yet amid the pronouncements of churchmen designed to
justify the newly established order lurked their uncertainty
about its survival. Their ambiguity and discontent, so evident
at the time of the Revolution, made them suspicious of ten-
dencies at work in this new constitutional and political structure. It is possible for a nation to ignore God's design and the
signs by which he reveals it, and if men defy providence God
"fit[s] them for Destruction."[71] Providence is ever-operative
in creation: "when God had made the World, he did not leave
it to shift for itself, without any farther regard of it. But his

[68] Anon., *A Resolution of Certain Queries Concerning Submission
to the Present Government* in *A Collection of State Tracts* (London,
1705), I, 442; cf. Fleetwood, esp. 46.

[69] Fleetwood, 100, Jan. 30, 1698/9.

[70] William Lloyd, *A Discourse of God's Ways of Disposing of
Kingdoms*, pt. I (London, 1691), 25.

[71] W. Lloyd, *A Sermon Preach'd before the House of Lords . . .
on 30th January, 1696/7* (London, 1697), 20.

Power does as truly appear in the Preservation and Government thereof, as it did in its Creation."[72] The Revolution of
1688–1689 was part of God's plan, indeed "the marks of
God's Hand were so visible in it, at first, and are so daily more
and more; that he is blind that doth not see them. There is
enough, one would think, to convince even the Atheist to the
belief of a Providence. . . . It is plainly the design of God . . .
to establish the Protestant religion in these Kingdoms."[73] The
Reformation will be accomplished only if men recognize and
act in accordance with the providential plan. Yet "the un-
thinking sort of Men . . . are . . . the greatest part of the
Body of a Nation. And when all these go together, they are
like the Atoms of Air, which though taken apart they are too
light to be felt, yet being gather'd into a Wind, they are too
strong to be withstood."[74] The instincts of the majority must
be curbed; they must be made to see the outline of the providential plan.

To bring the Reformation to fulfillment, to accomplish the
providential plan, churchmen assailed vice and corruption.
With increasing vehemence after 1688–1689 churchmen attacked the vices of the age: lewdness, swearing, drunkenness.
They admonished the Restoration period for its looseness and
even praised the Interregnum as a time when vice and immorality were checked.[75] This puritan reformism led to the
creation of societies for the reformation of manners. Ostensibly
they were intended as instruments in the creation of a new
moral tone preparatory for the general reformation. They
were more expressive, however, of a feeling of political and
social impotence common in church circles after the Revolution. With the lessening of the church's power, it was felt,

[72] John Moore, "Two Sermons before the Queen, August 17th and
24th, 1690," in Clarke, ed., 122–123.

[73] W. Lloyd, *A Sermon Preached before the Queen . . . January
30th, 1690/1* (London, 1691), 29.

[74] Lloyd, *Discourse*, pt. I, 14.

[75] Lloyd, *Sermon . . . January 30th, 1690/1*, 24.

came the unleashing of a torrent of impiety.[76] These societies
gained support largely from the moderate wing of the church,
clerics such as Josiah Woodward, the Boyle lecturer, Thomas
Tenison, and William Lloyd; and from certain prominent
Dissenters such as Edmund Calamy. Shortly after the societies'
formation, worldly minded figures such as James Vernon saw
them as a form of hypocrisy,[77] and by the middle of Anne's
reign they were attacked for subverting the Anglican establishment or for suppressing and persecuting the poor.[78] More
conservative bishops such as John Sharp had reservations
about the absence of clerical control in some of the societies.[79]

But to Lloyd and many others, the vices that the societies
aimed to control were a portent of the final destruction awaiting a sinful nation.[80] Alongside the development of a social
and political ideology centering on the idea of God's providence and on the order inherent in God's governing of the
universe existed fears and anxieties about the possible decay
and destruction of the order prescribed in society and nature.
In a sermon to Queen Mary, Tenison connects the concern
over vice and immorality with this fear of destruction: "He
only who is not conscious to himself of guilt unrepented of,
who is sincere in his duty to God and man, is capable of remaining without despair, if he could perceive the frame of
Nature ready to be dissolv'd."[81] Tenison was a trustee of the
Boyle lectureship and thereby became one of the promoters
of the Newtonian natural philosophy with its emphasis on
order and regularity in nature. At the same time he was

[76] Dudley W. R. Bahlman, *The Moral Revolution of 1688* (New
Haven, 1957), 27.
[77] George P. R. James, ed., *Letters Illustrative of the Reign of William III from 1696 to 1708* (London, 1841), II, 133, July 21, 1698.
[78] Bahlman, 83–87.
[79] Sharp MSS, Borthwick Institute, York, microfilm from Box 4/T,
Worcestershire Record Office.
[80] Lloyd, *Sermon . . . January 30, 1691/2,* 31.
[81] Tenison, *A Sermon Concerning the Folly of Atheism . . . , February 22, 1690/1* (London, 1691), 10.

preaching the possible destruction of the world natural. He and many of his latitudinarian colleagues accepted that frightening possibility because they believed that in the providential plan the destiny of the natural order hinged on the course taken by the social and political order. If the church lost power and influence and "crafty, ill-principled" men, as Evelyn called them, came to dominate the private, and most important, the public life of the nation, they would undermine the Protestant Reformation and cast their will in opposition to the providential plan. As punishment a wrathful God would wreak destruction upon both the natural and the political orders.

Latitudinarian churchmen accepted the Revolution of 1688–1689 and indeed sought to justify it. Yet doubts and uncertainties remained and were translated into renewed efforts to promote natural religion as a means to reconcile all Protestants. The acceptance of a broad and liberal Protestantism would make possible the fulfillment of the Reformation and the accomplishment of the providential plan. In anxious anticipation of that event some churchmen sought knowledge of the past which would assist them in predicting the future. They searched the Scriptures and commentaries for a hidden key that would reveal the precise stages through which history must pass in the course of arriving at its predetermined destiny. The end of history, the culmination of the "world politick" and the world natural, promised victory for the church and those who had served it loyally. Within the latitudinarian faction of the church the belief persisted that ultimate victory entailed the establishment of a new, millennial order in nature and society. They believed that Christ would come again and destroy the worlds natural and politick and on their ruins establish a millennial paradise, a new order in nature and society more regular and ordered, more just and perfect, than anything previously experienced or imagined.

CHAPTER 3

The Millennium

The latitudinarian faction of the church, as well as certain Newtonians, assumed and explicitly discussed the coming of the millenarian paradise. It would fulfill the providential plan revealed in Scripture, the historical dimensions of which lay buried amid the cryptic visions and prophecies of Daniel and St. John. The new heaven and the new earth would be built, both physically and morally, on the ashes of the old political and natural order. So certain were church millenarians of the course intended by God—the promise of a church-dominated paradise—that they imagined only one possibility capable of undoing the divine plan. True to their interpretation of free will, churchmen believed that only human sinfulness, repeated and unrepented, could thwart providence. Flagrant disobedience of the dictates of the providential order by the nation in its public and political dealings would anger a watchful deity who in retribution would destroy both the political and the natural order and all who dwelt therein.

Due warning would precede this literal and physical destruction. Churchmen thought that various devices might be employed: an earthquake (one had occurred in England in 1692),[1] a comet crashing into the earth, or an imbalance in

---

[1] Edmund Calamy, *An Historical Account of My Own Life with Some Reflection on the Times I Have Lived in (1621–1731)* (London, 1829), I, 326. See also J. E. Foster, ed., *The Diary of Samuel Newton, Alderman of Cambridge (1662–1717)* (Cambridge, 1890), 106; and Evelyn to Tenison, in John Evelyn, *The Diary and Correspondence* (Bohn ed., London, 1859), III, 325–330. The original manuscript ver-

gravitational pull causing the planets to crash into one another. Newton may have been providing a natural mechanism for such an event, should it be needed, when he spoke, in the thirty-first query to the *Opticks* (1717–1718), of the "inconsiderable irregularities . . . which may have arisen from the mutual actions of comets and planets upon one another, and which will be apt to increase till this system wants a reformation."[2] Natural mechanisms for the destruction of the "world politick" and the world natural could also be employed by God as the means by which the millennium would be instituted. The reason for the destruction of the natural world, revenge for human sinfulness or the fulfillment of the historical scheme preordained by providence, would depend solely on man's responsiveness to divine will.

At times of political crisis some churchmen pondered events as they occurred, both in politics and nature, for the correspondence of those events to the providential plan. Indeed throughout the seventeenth century in England millenarian speculation, whether among churchmen or radical sectaries, arose as an almost standard response to political and social instability and upheaval. We have seen that in 1688, Evelyn interpreted earthquakes in Smyrna as possible signs of God's displeasure with the impending revolution; at the same time he drafted a cautiously optimistic treatise concerning the millennium. The English were, after all, the chosen people, and

---

sion of this letter is clearer and more easily understood, MS 931, f. 59, Lambeth Palace Library.

[2] Newton, *Opticks: or a Treatise of the Reflections, Refractions, Inflections and Colours of Light* (2d ed., with additions, London, 1718), 378. Similarly Newton thought that "the variety of Motion which we find in the World is always decreasing" (375). This raises the whole question of the role of active principles in Newton's thought. See Henry Guerlac, *Newton et Epicure, Conférence donnée au Palais de la Découverte le 2 Mars 1963* (Paris, 1963), and J. E. McGuire, "Transmutation and Immutability: Newton's Doctrine of Physical Qualities," *Ambix*, 14 (1967), 69–95, and D. Kubrin, "Newton and the Cyclical Cosmos: Providence and the Mechanical Philosophy," *Journal of the History of Ideas*, 28 (1967), 325–346.

natural events, wherever manifested, might be intended for their edification.

For the most part, however, the latitudinarians believed that providence operated in nature not by recourse to the dramatic and portentous such as earthquakes or crashing comets, but through the imposition or order and harmony. In God's work design is everywhere apparent and purposeful; nature works for man's benefit so that in turn he will conform his desires and interests to the providential plan. His word revealed in Scripture prescribes the means by which this plan must be fulfilled. If read carefully, Scripture records the unfolding of that plan in history, in past events and in present and future happenings. Knowledge of the divine word, in particular the prophetic texts, will enable the church to guide events in conformity with the providential will, in expectation of the millennial paradise.

Anglican churchmen who indulged in millenarian speculations ascribed to a practice that permeates the Christian tradition. For centuries Christians had found in the prophetic texts, especially in the books of Daniel and St. John, justification for their historical actions. To the victims of history, the poor and oppressed, or to the opponents of historic orthodoxies, the heretical and rebellious, the Scriptural prophecies offered hope or even justification for rebellion. They read them as vindication of their dreams; the prophecies foretold a righteous order coming to pass, an earthly paradise to be established by Christ at his second coming wherein his followers, the saints, would reign triumphant. In the ideologies of the oppressed and their radical movements which span the medieval and early modern periods, the millennial prophecies provided a utopia.[3] The pursuit of the millennium justified the destruction of a pernicious political order ruled by beasts and whores of Babylon. Early in the Reformation the ominous identification was made between the Roman church and antichrist. Luther accepted it,

[3] Norman Cohn, *The Pursuit of the Millennium* (London, 1970); Karl Mannheim, *Ideology and Utopia* (London, 1936), 190 et seq.

but Calvin shied away. The divine plan as revealed in Scripture possessed allegorical significance, but Calvin's God and therefore his plans for the future remained inscrutable.[4]

The tenets of sixteenth-century Calvinism, however, bore less and less resemblance to Calvin's original intentions, and among English Puritans the identification of Rome with antichrist became a part of standard diatribe. Millenarianism in early modern Europe often accompanied social and political radicalism as witnessed by the Anabaptists at Münster or the radical sects of the English Commonwealth.[5] But the linkage between millenarianism and political radicalism was separable. Political moderates in seventeenth-century England, indeed even political and social conservatives, could and did believe, sometimes tentatively and cautiously, that the Scriptural prophecies proclaimed the advent of a millennial paradise that drew near on a predetermined and discernible time scale.[6] The espousal of millenarian doctrines by social conservatives surprises us only because we have mistakenly and exclusively identified those beliefs with the radicals and because by and large the historical evidence for the latter has been more apparent.

Yet how could Anglican churchmen so late in the seventeenth century still find comfort in millenarian doctrines associated with mid-century radicals and discredited, at least in established circles, along with their radical causes? During the 1650s the radicals had threatened to turn the world upside down and to destroy the very social and political order that maintained the church.[7] After 1660 the restored church bent

[4] William Lamont, *Godly Rule, Politics and Religion 1603–60* (London, 1969), 22–23.

[5] Cf. Cohn, 252 et seq., and appendix.

[6] William Lamont, "Richard Baxter, the Apocalypse and the Mad Major," *Past and Present*, No. 55 (1972), esp. 74, and *Godly Rule*, 56–77; cf. Bernard Capp, "*Godly Rule* and Millenarianism," *Past and Present*, No. 52 (1971), 106–117. I obviously agree with Lamont on the question of "optimism" and "pessimism."

[7] The best discussion of the radicals occurs in Christopher Hill, *The World Turned Upside Down* (London, 1972).

104   The Newtonians and the English Revolution

every effort to forestall the possibility of such an upheaval ever happening again. During the Restoration millenarian sentiments were almost never expressed in church sermons, although as mentioned at the end of Chapter 1, it is possible to find occasionally such sentiments expressed by Tillotson and Tenison and possibly by others. Much more needs to be known about conservative millenarianism in that period. If Richard Baxter and Thomas Hobbes can be classified as millenarians[8]—albeit as very profoundly different ones—my suspicion is that much more could be said on this topic, that a story could be told about the church and the millennium throughout the seventeenth century. My concern here is only for the period after the mid-1680s.

The millenarianism of late-century church moderates and certain Newtonians attests to the all-pervasive character of such speculation within English Protestantism and more particularly to the protean character of millenarianism. This Anglican millennium differed quite purposefully from that of the radicals. It would place the church, and not simply the saints, as triumphant in the "new heaven and the new earth." In the millenarian paradise and consequently in the historical process that would lead to its creation, the church and the more powerful and prominent members of its laity would lead the nation and finally the world along the stable and peaceful course preordained and guided by providence. The Protestant Reformation with the English church as its vanguard would convert both the Old and the New World to its teachings, and the antichrist, Rome and more immediately its main political defender, the French king, would suffer defeat and finally destruction at the hands of Protestant forces. Furthermore, Anglican millenaries like Evelyn and Thomas Burnet insisted first on the destruction of this earth by Christ and the establishment of an earthly, but new kingdom. The

[8] Lamont, "Baxter," 68–90; J. G. A. Pocock, "Time, History and Eschatology in the Thought of Thomas Hobbes," in J. H. Elliott and H. G. Koenigsberger, eds., *The Diversity of History* (Ithaca, 1970).

social message in this sequence is obvious: the radicals could never establish a millenarian kingdom in this world according to their social principles in anticipation of the second coming. Their millennium contradicts Scripture; yet Scripture clearly predicts the advent of the millennial paradise. In direct contradiction to the challenge to that order posed by the radical millenarians, the latitudinarians sought the attainment of their paradise through the acceptance of a broad and tolerant Protestantism that rewarded private endeavor and at the same time maintained the church's vested interests. The millennial paradise would not come until Protestantism had been rescued in England, in Europe, and finally in all the known world.

Whenever the advance of Protestantism was endangered, as it was in the 1680s, churchmen in England, like other European Protestants, resorted to apocalyptic speculations. The advent of a Catholic monarch in England and the dispersion of French Huguenots as a result of the revocation of the Edict of Nantes, both in 1685, combined with Louis XIV's aggressive foreign policy to create an almost hysterical atmosphere in some Protestant quarters. Among French Huguenot refugees, Pierre Jurieu denounced the French king as the incarnation of the beast and leader of the doomed Roman Empire. Jurieu and his followers advocated a holy war that would complete the Reformation and accomplish the Scriptural prophecies.[9] Exiled in Holland, they looked to William of Orange as the last hope for European Protestantism, and they supported the Orangist cause with complete dedication. The Huguenot exiles were probably not the only millenarian group in The Netherlands in the 1680s, but little is known about such groups, if still they existed.[10] Within France millenarianism reappeared in Languedoc in the 1690s and became an inherent element in

[9] See W. Rex, *Essays on Pierre Bayle and Religious Controversy* (The Hague, 1965), 197–255; also G. H. Dodge, *The Political Theory of the Huguenots of the Dispersion* (New York, 1947); P. Jurieu, *The Accomplishment of the Scripture Prophecies* (London, 1687).

[10] For evidence of Dutch millenarianism in the 1660s see B. S. Capp, *The Fifth Monarchy Men* (London, 1972), 197, 214.

Camisard opposition to the crown. The millenarianism of the
Camisards resembles, however, the style, if not the doctrines,
of the radical reformation, and when the French prophets
made their way to England in 1705–1706 even millenarian
churchmen recoiled from them.[11]

In England during the 1680s the prophecies aroused some
popular interest. Remnants of the radical sects of the civil
wars, such as the Fifth Monarchists, rallied around the cause
of the duke of Monmouth.[12] Among the Baptists, Benjamin
Keach warned that the prophecies were near fulfillment and
Hanserd Knolly's *An Exposition of the Whole Book of the
Revelation* (1668) was reprinted with a license dated September 12, 1688. In the same year two of Jurieu's works appeared in English.[13] In 1689, Thomas Beverley proclaimed
1697 as the year when the Reformation would be reformed
and John Mason added his prognostication about the coming
cataclysm.[14] Much of this prophetic excitement was provoked
by the growing opposition to James II's policies, the rumors
of an intended invasion from Holland, and finally the accomplishment of the Revolution.

In the Anglican church during the Restoration speculation
about the Scriptural prophecies and the time scale ordained
by providence for their fulfillment emanated most commonly

[11] See Chapter 7.
[12] Capp, 221.
[13] Jurieu, *A New Systeme of the Apocalypse* (London, 1688); *A
Continuation of the Accomplishment of the Scripture Prophecies*
(London, 1688).
[14] T. Beverley, *The Kingdom of Jesus Christ* . . . (London, 1689);
and *To the High Court of Parliament. Assembled M. 4. D. 30. 1691.
The Most Humble Address of T. Beverley* (London, 1691); Christopher Hill, "John Mason and the End of the World," in *Puritanism
and Revolution* (London, 1958), 323–326. See also Matthew Mead,
The Vision of the Wheels Seen by the Prophet Ezekiel (London,
1689). I do not think that millenarianism died a sudden death in
church circles after 1660. For the opposite view see Christopher Hill,
Anti-Christ in Seventeenth Century England (London, 1971), 154 et
seq. See also H. Maurice, *An Impartial Account of Mr. John Mason
of Water Stratford* (London, 1695), cited in Hill, *World*, 160.

from Cambridge. Henry More and John Worthington were both social conservatives and millenarians who attacked the radicals and stole their fire by proclaiming a millenarian paradise firmly controlled by the church.[15] More proclaimed that the Reformation initiated by Luther marked the beginning of the Roman Empire's fall, and "the most certain sign of the downfall of anti-Christ will be the raising again in the Reformed Churches a sincere and fervent Zeal after Truth and Holiness, hearty Love and Amity among themselves."[16] Much of this Cambridge-based speculation rested on the earlier writings of Joseph Mede and more distantly of the Elizabethan mathematician John Napier.[17] It is more than possible that the millenarian speculation infusing latitudinarian circles owed its origin directly to the teachings and writings of the Cambridge Platonists.

Certainly one of the most important statements of Anglican millenarianism in the late 1680s, Thomas Burnet's *Sacred Theory of the Earth*, must owe some inspiration to the author's direct association with More and Cudworth. Burnet went up to Cambridge in 1651 and resided at Clare Hall until he followed Ralph Cudworth to Christ's. Throughout the 1660s, Burnet remained associated with the college, and his close relationship with Cudworth inevitably meant his exposure to the thought of Henry More, a long-standing fellow of Christ's.

[15] John Worthington, *Miscellanies . . . Observations Concerning the Millennium . . .* , ed. E. Fowler (London, 1704). The latitudinarians took considerable interest in the work of Matthew Poole; see M. Poole, *Annotations upon the Holy Bible . . .* [a continuation of Poole's work by S. Clark and E. Veale] (London, 1700), II, Rev. 16: 10.

[16] Henry More, *Apocalypsis Apocalypseos* (London, 1680), 25.

[17] Joseph Mede, *Clavis Apocalyptica* (London, 1627), trans. R. B. Cooper, 1833; John Napier, *A Plaine Discovery of the Whole Revelation of St. John . . .* (London, 1593). See D. Castillejo, *A Theory of Shifting Relationships in Knowledge* (n.p., 1968), III, 560, copy in U.L.C.; and A. Gilsdorf, "The Puritan Apocalypse: New England Eschatology in the Seventeenth Century" (Ph.D. diss., Yale University, 1965).

There are other possible Anglican sources for Burnet's millenarianism. Drue Cressener (1638?–1718) indulged in apocalyptic speculations in the 1680s, if not before, and he was at Christ's with Burnet. In 1662 he removed to a fellowship at Pembroke Hall and remained active in Cambridge affairs throughout the Restoration. Cressener was on good terms with William Lloyd, bishop of St. Asaph, that most avid millenarian, and with Simon Patrick who supported Cressener's endeavors.[18] Simon's brother, John, also appears to have been sympathetic to millenarianism,[19] as were Archbishop Sancroft and John Evelyn. Edward Fowler, John Tillotson, and Thomas Tenison,[20] all Cambridge men, express apocalyptic sentiments in their sermons, and Fowler was so gripped by millenarian fervor and possibly by senility (he was born in 1632) that during the reign of Anne his name appears on the list of followers or sympathizers drawn up by the French prophets.[21] Burnet could have drawn on any of these sources; indeed on the basis of available evidence we can be confident that private millenarian speculation was commonplace in certain church circles.

Domination by a Catholic monarch and Protestant opposition to his policies broke down the reticence of some churchmen to preach openly about the millennium. In 1689, Thomas Burnet published his last two books of the *Sacred Theory* in

[18] See letters printed in D. Cressener, *The Judgements of God upon the Roman Catholick Church . . . with a Prospect of These Near Approaching Revolutions. . . . In Explication of the Trumpets and Vials in the Apocalypse . . .* (London, 1689), and preface to *A Demonstration of the First Principles of the Protestant Applications of the Apocalypse* (London, 1690).

[19] Tanner MSS XXVI, f. 44, Bodleian.

[20] See concluding remarks in Chapter 1; T. Tenison, *A Friendly Debate between a Roman Catholick and a Protestant, Concerning the . . . Transubstantiation . . . and Anti-Christ Is Clearly and Fully Described . . . September 24, 1688* (London, 1689?).

[21] Fatio de Duillier MSS, MS français 603, Bibliothèque Publique et Universitaire de Genève. Fowler's letters to Locke in 1704 indicate poor health, but hardly senility; MS Locke, c. 8, Bodleian.

Latin and openly proclaimed his millenarianism and its relevance to the church's predicament. By the time the book
appeared, however, the church's hegemony had been partially restored by the Revolution and Burnet's intention in
writing the *Sacred Theory* was obscured. In the subsequent
English edition published in 1690, Burnet quietly toned down,
indeed almost obliterated, his millenarian ardor. The Revolution removed the immediate occasion for its expression, and
since millenarianism appeared superfluous or dangerous to the
domestic situation Burnet significantly changed his text. The
Latin edition was forgotten or assumed to be exactly similar to
its supposed English translation, and consequently Burnet
has commanded attention primarily as one of the first natural
historians. His *Sacred Theory* becomes simply a first attempt,
in modern times, to devise a cosmogony and a natural history
of the earth.

Published in two parts—the Latin part I of 1681 translated
into English in 1684 and the Latin part II published in 1689
and translated in 1690—the *Sacred Theory* begins with an
account of the earth's origin, formation, and present condition,
and ends with a description of its final destruction and the
new heaven and new earth which will be built on its ruins.
Throughout this account Burnet supposes that he is supple-
menting the biblical version of the earth's natural history.
Largely ignorant of other geological speculation in the seventeenth century, Burnet relies primarily on Cartesian explanations of the forces at work in the earth's formation. He justifies his efforts by claiming that the Mosaic account has been
left incomplete purposefully and that the time has come for
someone to augment the story. Thus Burnet says that under
the guidance of divine providence the flood occurred as a
result of the pressure that cracked the earth's shell and forced
the trapped water to inundate the surface; he also says that
the final conflagration will begin as the fire in the bowels of
the earth creates the pressure necessary for a fiery explosion.
Such a combination of geological theorizing and biblical

exegesis strikes the modern reader as curious, and at least one commentator has attempted to explain this combination by arguing that "both intellectually and chronologically Burnet was a scientist and scholar before he was a theologian."[22]

But if an assessment of the religious and political situation at the time Burnet wrote the *Sacred Theory* is combined with a careful reading of both the Latin and English editions of part II of the *Theory*, the seeming dichotomy in Burnet's thought disappears. My research with important textual assistance from Wilfrid Lockwood has produced convincing evidence that Burnet used his understanding of natural agencies, which was limited, and his techniques in biblical exegesis, not as a scientific exercise loosely supported by Scripture, but to present a plausible explanation of how creation occurred and, more important for his time, of how the millennium would be instituted—all in accordance with the Scriptural account. To Burnet this explanation was of vital importance because he intended to give credibility to the millenarian prophecies at a time when these prophecies bore special relation to political events, both in England and on the Continent. Burnet identified his interests entirely with those of the church; when he wrote part II of the *Sacred Theory* he held the influential ecclesiastical position of master of Charterhouse. Like many other Protestants, Burnet believed that the struggles against the domination and power of Rome were foretold in the Scriptural prophecies. Unlike many of his church colleagues,

[22] Katherine B. Collier, *Cosmogonies of Our Fathers . . .* (New York, 1934), 69. For a discussion of Burnet's place in the history of science, see P. Rossi, *Aspetti della rivoluzione scientifica* (Naples, 1971); R. Lenoble, *La geologie au milieu de XVIIe siècle* (Paris, 1954); and John C. Green, *The Death of Adam* (New York, 1961). The work of David Kubrin deserves thoughtful consideration: "Providence and the Mechanical Philosophy: The Creation and Dissolution of the World in Newtonian Thought" (Ph.D. diss., Cornell University, 1968). In one account, G. L. Davies, *The Earth in Decay* (London, 1969), 68–74, the last two books of the *Theory* are dismissed as "fancies."

Burnet had openly opposed James II and blocked his attempt to install a Catholic pensioner at Charterhouse.

Burnet was not alone, however, in his attempt to proclaim the relevance of the Scriptural prophecies to events. In 1689, Drue Cressener published a work entitled *The Judgements of God upon the Roman-Catholick Church . . . with a prospect of these near Approaching revolutions. In explication of the trumpets and vials of the Apocalypse, upon principles generally acknowledged by Protestant interpreters.* Cressener begins by stating that his book (up to the nineteenth chapter) had been completed by March 1688 and was in circulation when the bishops were sent to the Tower in June. To verify his claim, Cressener published signed statements from Simon Patrick, then bishop of Chichester, Henry Plumptree, and Thomas Burnet testifying that they had read Cressener's work in the spring of 1688. In that year these Anglicans formed a circle with avid interests in the millenarian prophecies.

Why were Cressener and his friends so concerned to prove the date when his treatise had been written? In it he argues that the resurrection of the witnesses and the recovery of the Protestant churches will occur in 1689 or at the latest 1690. Obviously, in the wake of William's successful invasion and the subsequent restoration of Protestantism, Cressener did not want his prophecy to be written off as mere hindsight. Cressener does not claim that the millennium will begin in 1689; on the contrary, the fifth, sixth, and seventh vials must first be poured out. This process, all a part of the third woe, will culminate in the destruction of antichrist. It will be swift, for the tide has turned against Babylon. Although Cressener submitted his book for approval to Lambeth Palace shortly after it was written, for some reason its publication was delayed. Perhaps the predictions made in it, based on the authority of Scripture, could have been seen as an intended justification of rebellion, and in the summer of 1688 the church's hierarchy, with one or two exceptions, opposed such a venture.

Cressener's book was not the only publication that was delayed in that year. Like it, part II of Burnet's *Sacred Theory* was circulated privately; indeed Cressener, as well as Evelyn, saw a prepublication copy.[23] But publication of the *Sacred Theory* was held up for a time by ecclesiastical officials in Lambeth Palace. Writing in July 1688 to Robert Southwell, Burnet complains, "I have finished ye 2 remaining bookes of ye Theory of ye Earth, which you are pleased to enquire after, and ye papers have been at Lambeth 2 months for a license, and are not yet dismisst. You will think them too nice and scrupulous in such a conjuncture of time as this is, when all lyes at stake, to quarrel their friends about speculations."[24] At the time of Burnet's letter, schemes had been hatched for William's intended invasion, and the political situation as well as the position of the church had been precarious and unstable since the spring.[25] Despite this situation, Burnet did eventually receive permission and in September he reports to Southwell that his book has passed "ye pikes" at Lambeth, and that it concerns "ye conflagration of ye world, and ye new heavens and Earth that are to succeed."[26]

Burnet's acquaintance with Henry More and later with Drue Cressener reveals some of the possible sources for his

[23] Evelyn MS 35, Christ Church, Oxford. The *Sacred Theory* was not published before December 1688. It bears the date 1689. Evelyn's treatise was written in 1688 and makes no mention of the Revolution. It is most unlikely that it was written as late as December. In *The Judgements* . . . , 288, Cressener states that he finished chapter 19 and then saw the *Sacred Theory*, pt. II, sometime in the spring. Burnet finished it in May, if not before. Hooke lectured before the Royal Society on pt. II of the *Sacred Theory* in December or January. R. W. T. Gunther, *Early Science in Oxford*, X, *The Diary of Robert Hooke* (Oxford, 1935), 75, 81, 89.

[24] MSS ADD 10039, f. 63, July 10, B.L.

[25] John Carswell, *The Descent on England* (London, 1969), chap. 10.

[26] MSS ADD 28104, f. 16, Sept. 25, B.L. In this context read "pickers" for "pikes."

own millenarianism. Certainly it is stated quite openly in his posthumous works, such as *De Statu Mortuorum et Resurgentium* (London, 1728).[27] The tradition of Anglican millenarianism to which Burnet was exposed, coupled with the political crisis that provoked renewed interest in the prophecies, provides the historical context within which Burnet's *Sacred Theory*, in particular the last two books, can be interpreted. At the time when he wrote, Burnet was immersed in political affairs, particularly as they related to James's attempts to install a Catholic pensioner at Charterhouse. As well as firmly opposing these encroachments on church prerogatives, Burnet supported the parliamentary candidacy of men like Southwell whose opposition to royal policies was well known.[28]

How then, precisely, does the *Sacred Theory* relate to these political events and to the tradition of millenarianism found in church circles? In the *Sacred Theory*, Burnet attempts to add another dimension to the struggle against antichrist. The very course of nature, the cosmology of the universe, depends for its fulfillment on the defeat of the beast. Relying on the Scriptural account of human history, Burnet attempts to provide a plausible explanation of exactly how, in physical terms, the millennium will occur. The biblical prophets, in particular St. John, had foretold the stages through which mankind would pass en route to the millennial paradise. Guided by divine providence nature also follows this preordained course that culminates in the creation of a physical paradise. That paradise stands apart from the present condition of the earth, which will be destroyed by conflagration. Burnet rejects the possibility of any future improvement of man's earthly condition. He was dissatisfied with our world; neither its natural nor its moral history offered consolation. Even the physical

[27] *De Statu*, 101–120; pt. II, 16–17.
[28] T. Burnet, *A Relation of the Proceedings at Charter-House, upon Occasion of King James II His presenting a Papist* . . . (London, n.d.); MSS ADD 28104, f. 16, B.L.

contours of the earth, its mountains and valleys, were aesthetically unappealing.[29] Hope lay in the fulfillment of the Scriptural prophecies, in the destruction of the beast, and the subsequent creation of a new heaven and a new earth built upon
the ruins of the present physical and political order.

The evidence for interpreting the *Sacred Theory* in a way
that relates it both to the events of 1688–1689 and to the political millenarianism of Anglican thinkers comes from a careful
reading of the Latin edition of the *Sacred Theory*, part II,
written before the Revolution (although published in 1689),[30]
and from a comparison of this text with the English translation of 1690. Contrary to the general assumption that the
English text is merely a translation of the Latin, a careful
comparison of the texts shows that the English version is very
far from being a straightforward translation. It is true that the
general order of the Latin is preserved, and there is close correspondence between extensive passages. However, much has
been altered, sometimes slightly, sometimes drastically, and a
great deal of the work has been totally rewritten. It will be
useful, first, to indicate one important feature shared by both
versions. This is the general theory that the millennium will
take place on this earth; but since this present earth is mani-

[29] Although Tuveson recognizes the existence of Anglican millenarianism, I cannot agree with his interpretation of Burnet; see E. Tuveson, *Millennium and Utopia: A Study in the Background of the Idea
of Progress* (New York, 1964). On Burnet's aesthetics see M. H. Nicolson, *Mountain Gloom and Mountain Glory* (Ithaca, 1959). Of related interest is Sylvia L. Thrupp, ed., *Millennial Dreams in Action*
(The Hague, 1962).

[30] E. Arber, *The Term Catalogue* (London, 1906), II, 239, possibly
in December. Burnet must have written the dedication to James, duke
of Ormond, in late July 1688, when his dukedom was confirmed. It is
only reasonable to suppose that the English translation was begun after the publication of the Latin, especially since the Latin copy in
Cambridge University Library, Burnet's own, indicates that he probably projected another Latin edition before finally undertaking the revised English one. I am most grateful to Wilfrid Lockwood of the
Cambridge University Library for his translations and assistance with
these texts.

festly misshapen and imperfect, the millennium must follow the total destruction by fire of the present face of nature and its renewal in a form more aesthetically pleasing. Indeed, what is distinctive about Burnet's thesis is that the same natural or secondary causes which work in the premillennial world, for example, fire, serve to bring in the millennial order. Hence, Burnet's main preoccupation is his attempt to combine scientific reasoning and millennial prophecy. Nevertheless, the *Sacred Theory* is a millenarian document, though much more obviously so in the Latin version than in the English.

Since I and Wilfrid Lockwood have published elsewhere an extensive comparison of the Latin and English texts,[31] it seems sufficient here to summarize the differences between them. They consist mainly of the omission in the English text of 1690 of passages with a markedly millenarian character. Yet there are other interesting and revealing differences between what Burnet said on the eve of the Revolution and what he said after it. In the Latin version he speaks of justice as "hodierna (terra) exul"—an exile from the earth today—and he speaks despondently about the condition of the saints in their present life as contrasted with the future happiness promised them in the millennium.[32] After the Revolution, Burnet removed his bitter references to persecution and injustice; he contented himself with condemning worldly men who seek pleasure and preferment.

The most intriguing problem presented by Burnet's millenarianism is to discover the time scale he envisioned for the coming of the millennium and the destruction of this earth. Burnet, unlike the radicals, would have destruction first and then the earthly paradise. As he explained: "Our opinion hath this Advantage above others, that all fanatical Pretensions to Power and Empire in this World, are, by these Means, blown

[31] M. C. Jacob and W. A. Lockwood, "Political Millenarianism and Burnet's *Sacred Theory*," *Science Studies*, 2 (1972), 265–279.
[32] Burnet, *Telluris Theoria Sacra . . . Libri Duo Posteriores de Conflagratione Mundi, et de Futuro Rerum Statu* (London, 1689), 3–4, 197, 203.

away, as Chaff before the Wind. Princes need not fear to be dethroned, to make way to the Saints, nor Governments un-king'd that they may rule the World with a Rod of Iron."[33] Another advantage of Burnet's theory, he claims, is that no one's material circumstances need be altered in the millennial paradise[34]—in other words, property and income distribution remain untouched. This delightful eventuality for the few will occur within one or two centuries,[35] probably at the end of the nineteenth century.

Other church millenaries would have disagreed with Burnet's time scale; William Whiston wanted it earlier as did William Lloyd. Newton, on the other hand, probably accepted the year 2000. One of the major differences between churchmen and radicals was the urgency of their respective millenniums. The more enthusiastic radicals wanted it soon, often within a year or two after they proclaimed it. Churchmen in the late 1680s felt that they could afford to wait until a church-dominated Protestantism had been rendered secure.

In the English version of the Sacred Theory, Burnet never even hints that he has worked out a time scale. He makes vague references to a future paradise, a new heaven and a new earth, but they are so unclear that modern readers have never realized that Burnet was a full-fledged millenarian. His interest in natural history has aroused scholarly attention, but, of course, that interest sprang from his desire to synchronize the operations of history, of the "world politick," with the world natural. That attempt led Burnet into an inevitable trap: how to allow for direct, divine intervention in both history and nature, since Christ would have to come again, and still retain the regular and ordered universe of the new science, governed by providence through laws or secondary causes.

[33] Quoted in Kubrin, "Providence," 104, from the 1726 review of Burnet's *Theory*, 396–397.

[34] Burnet, *De Statu Mortuorum et Resurgentium Tractatus* (2d ed., London, 1728), 101–120.

[35] Burnet, *Telluris Theoria Sacra*, 1689, dedicatory epistle, 240–241.

In Burnet's own private copy of the Latin *Sacred Theory*, deposited at the Cambridge University Library, we see him struggling with the problem presented by direct divine intervention. He would have Christ hovering over Rome and smiting "the seat of Anti-Christ with a burning ray from his mouth, and then, as if at an agreed signal, the whole of nature will at once burst into flame."[36] In his own hand he has crossed out "at once" and substituted, in Latin, "gradually, or after a short delay." But the problem of being a millenarian and yet a proponent of science never bothered Burnet or his other church colleagues sufficiently to lead them to abandon one or the other. Both enterprises—showing the order and stability of nature as governed by providence and the eventuality of the millennium—were too urgent, given the church's situation in an increasingly secularized social and political order, to warrant scruples over apparent intellectual contradictions. As Keith Thomas has noted, "The mechanical philosophy of the later seventeenth century was to subject this doctrine of special providences to a good deal of strain."[37] Yet these strains are seldom apparent in the writings, both private and public, of the latitudinarians. Their willingness to ignore these intellectual problems attests to the overriding importance of their ideological and political interests. They wrote and preached to a population steeped in the belief that providence operates in both the regular course and in special events of nature,[38] and the latitudinarians, like their listeners, could not divorce themselves from that basic assumption. Burnet knew that someday Rome would burn, and the mechanism that would begin the conflagration posed an interesting problem for someone with scientific interests, but ultimately the questions of "when" and "why" were more urgent for Anglican thinkers than was the question, "how."

Burnet's private notes, intended as part of the preparation

[36] Ibid., 82.
[37] Thomas, *Religion and the Decline of Magic* (London, 1971), 80.
[38] Ibid., 78–89.

for a second edition of the Latin text, which never appeared, do not explain the major changes Burnet made in the revised English text. Direct divine intervention troubled him; he also wanted more documentary evidence for his interpretation of Scripture, but basically he seemed content with his millenarianism. Yet in 1690 it had disappeared. The differences between these two versions of the *Sacred Theory* seem best explained by reference to the events of 1688–1689.

Certainly Burnet's letter to Southwell in July 1688 displays Burnet's concern that his book be published "in such a conjuncture of time as this is." For some of his fellow churchmen, such as Lloyd and Cressener, it was a time about which the Scriptural prophecies had spoken, when yet another step would be taken toward the destruction of this earth and the institution of the millennial paradise. In it the persecuted saints would reign triumphant and their centuries-old belief in the millennium would be vindicated. In the *Sacred Theory*, part II, Burnet attempts to add further evidence for this belief by offering an explanation of the earth's natural history that hinged entirely upon the predictions of Scripture. His opposition to the policies of James II and his reading of events on the Continent led him urgently to offer a millenarian interpretation of the earth's history. Written in haste, this Latin version of the *Sacred Theory* endeavored to awaken its readers to the truth of the millenarian vision and to bolster their courage in the face of antichrist.

Were Burnet and his circle using prophecy as their device to advocate a revolution? On other occasions in seventeenth-century England prophecy had certainly justified rebellion.[39] No evidence survives about the activities or response of Burnet, Cressener, and their clerical friends during the actual revolution of 1688–1689. In the summer and autumn of 1688, however, Burnet wrote revealing letters to Robert Southwell

[39] Ibid., 425. Cf. Anon., *Poor Robins' Prophecy, or, the Last Great News from the Stars: Foretelling the Mighty Wonders that Shall Happen in the Year 1689* (London, 1688).

bemoaning the treatment accorded the *Sacred Theory* at Lambeth Palace, and Burnet dedicated part II of the *Theory* to the young duke of Ormonde. Southwell, as Burnet had been,[40] was a loyal servant to the Ormonde household, and in October 1688 he wrote his master that shortly he would go to London and offer his services to the earl of Macclesfield. With candor seldom put in writing at the time, Southwell added: "I believe this nerve will be most of all amazing in [?] and that all our thunderbolts will light them. Besides what may fall from ye rest of Europe. They have a great desolation for humanity to account for, and it looks as if heaven were now disposed to send an avenger."[41] Southwell regarded the threatening invader as the very instrument of God, a sentiment which the duke of Ormonde may have shared. Certainly he was well into the Cockpit circle associated with the household of the prince and princess of Denmark and allied to the Orange plan for invasion.[42] Burnet's lay friends, Southwell and Ormonde, accepted the Orangist cause, and after the Revolution was accomplished Burnet was one of the first churchmen to preach before the prince.[43] It seems likely therefore that Burnet's sentiments were more Whiggish than those of many of his fellow churchmen; nevertheless the *Sacred Theory* could not be described as a revolutionary document. Its insistence upon the preordained course of events, which would inevitably bring destruction to antichrist, provides justification only for passive resistance. The very process of nature allies itself with the struggle against antichrist, and the beleaguered saint can

[40] Thomas Carte, *The Life of James Duke of Ormond . . . with an Appendix and a Collection of Letters . . .* (Oxford, 1856), IV, 698.

[41] MS 47A, Oct. 14, 1688, Forster Collection, Victoria and Albert Museum Library.

[42] David H. Hosford, "Bishop Compton and the Revolution of 1688," *Journal of Ecclesiastical History*, 23 (1972), 217–218.

[43] N. Luttrell, *A Brief Historical Relation of State Affairs* (Oxford, 1857), I, 490–492; on Ormonde, J. R. Jones, *The Revolution of 1688 in England* (London, 1972), 297. Burnet was in communication with the duke as late as November 1689; Burnet to Gascoine, Ormonde Correspondence, National Library of Ireland.

only be swept along by the preordained course of both human and natural history.

Once again, we have an example of a churchman, who despite his bitter opposition to James II's policies and the conspiratorial activities of his lay friends, dreaded any direct involvement in revolutionary activity. Burnet believed that somehow the same providence who guides both history and nature would save the Protestant cause and its true church, but its clergy could not take a direct part in aiding the work of providence. It was their task to explain and speculate about the providential order, even though in the hands of sectarian radicals those speculations could be extremely dangerous.

Since millenarian speculations were always open to misinterpretation by enthusiasts, churchmen, by and large, speculated in private. Some of the most interesting of those speculations are recorded in Evelyn's diary and his unpublished papers at Christ Church, Oxford. In April 1689, when the Revolution had been completed and William and Mary were firmly enthroned, Evelyn met with William Lloyd and William Sancroft, the still intransigent archbishop of Canterbury, to discuss the meaning of the Scriptural prophecies. Their able interpreter was William Lloyd whose interest in the mysteries of the Book of Revelation and the visions of Daniel had been long-standing. When Evelyn and Sancroft consulted Lloyd, he and Sancroft

entered into discourse concerning the final destruction of Antichrist: both of them concluding that the 3 Trumpet and Vial were now powering out; and my L.S. Asaph attributing the Killing of the two Witnesses, to be the utter destruction of the Cevenes [sic] Protestants, by the French and Duke of Savoy, and the other, the Waldenses and Pyrennean Christians (who by all appearance from good history had kept the Primitive faith from the very Apostles-time till now): The doubt His Grace Sancroft suggested, was whether it could be made evident that the present persecution had made so great a havock of those faithful people as of the other, and whether as yet, there were not yet some among them in being who met together: it being expedient from the Text:II:Apoc: that they should both be slain together: They both much approved of Mr. Mead's way of Interpretation, and that he onely failed in resolving too hastily, upon the King Swedens successes in Germany.[44]

With complete credulity, Sancroft and Lloyd accepted the meaning of the Book of Revelation as interpreted by Mede. The text portrays the stages or events leading up to the destruction of the beast and the second coming of Christ, the beginning of "a new heaven and a new earth." To them the persecution of the French Huguenots conformed to the events described in the second chapter of St. John's Revelations. Although Sancroft expresses certain hesitations based on the apparent lack of information about the extent of the persecution, Lloyd seems certain that recent events foretell the possibility of the final destruction of the antichrist. Significantly, this meeting took place in April 1689, some months after the English nation had been secured from the designs of a Catholic monarch.

Lloyd was not the only member of this gathering who had engaged in serious study of the Scriptural prophecies. A manuscript tract found among Evelyn's papers[45] reveals that in 1688 Evelyn was deeply concerned about the implications of the prophetic texts for the troubled events that foreshadowed the coming of William. The tract, "Concerning the Millennium," was written for and sent to the countess of Clarendon. It bears the date 1688 and, because Evelyn was in the habit of using new-style and old-style dates for days after January 1 and because internal evidence seems confirmative, the tract was written during 1688 and before the Revolution. He is at pains to make clear that his version of the new heaven and the new earth is "Not as the Millenaries of old, or Fifth Monarchists

[44] E. deBeer, ed., *The Diary of John Evelyn* (London, 1955), IV, 636.
[45] Evelyn MS 35, Christ Church, Oxford.

of Late, have fancied, for a Thousand yeares only before the final consumation; but [as I have said] after it [that], when our Blessed Saviour shall reign with his Saints forever, I do not say, on *this* earth at all; but in that Renewed Heaven and Earth to come."[46] Evelyn like Burnet insisted on the destruction of this earth by Christ and the establishment of an earthly but new kingdom. Evelyn tells the countess:

For this Earth in which dwell nothing but wickedness (as St. Peter tells us) shall be burnt up, purged and made another thing; the glorious state of which you will find described, through the whole Sixtieth Chapter of Isaiah. . . . And upon this Hope, the Apostle persuades them to take of that Refreshment, when Christ (whom the Heavens must receive in the meane time) shall be sent, as God had spoken by the mouths of his holy prophets since ye world began.[47]

Evelyn believes that only the saints, those who have repented their sinful ways and undergone conversion, will be spared during that final destruction in order that they may rule triumphantly in the new heaven and the new earth. In the best landowning tradition, "they shall build houses, and inhabit them" and enjoy "those earthly possessions of plenty and prosperity in the Land of promise."[48] In Evelyn's version of the saintly inheritance no reordering of existing economic inequities is ever hinted at. The saints progress toward spiritual and material fulfillment in the new heaven and the new earth proceeds gradually, and one essential step will be the freeing of God's people from the Tyrant Herod, the subsequent "revival of a purer Church," and the destruction of antichrist.[49] It is difficult to doubt that by naming Herod, Evelyn was referring to James II and to his hope for a restitution of the Protestant religion. Evelyn, of course, sought its recovery by any means except revolution. On November 2, 1688, he wrote

[46] Ibid., f. 2. Bracketed material crossed out by Evelyn.
[47] Ibid., f. 3.
[48] Ibid., f. 7.
[49] Ibid., ff. 7–8.

to Pepys about the possibility of James II's promoting Stillingfleet and Tillotson to bishoprics, presumably to stave off "the
impendent Revolution," and Evelyn assured Sancroft that he
had no use for invaders "be they Dutch or Irish."[50]

Despite his fear of Revolution, Evelyn, as we know, accepted the new order once the church's political influence had
been secured.[51] Evelyn's well-developed sense of self-interest
conspired with his millenarianism to render him a supporter of
the Revolution Settlement. In his role after 1691 as trustee of
the Boyle lectureship he became, until his death in 1706, one
of the most important guardians of church thinking and a
promoter of natural religion based upon Newtonian principles.
Among the motivations for Evelyn's actions, whether political
or intellectual, we must include his fervent belief in the approaching apocalypse. His extensive notations made in his
private Bible reveal Evelyn's scheme for the fulfillment of the
prophecies: "The 7th Viale poured on Antichrist was the
preaching of Luther, etc., and continues to this day: by all
the reformed; figured by the Earthquake . . . and that by
the Harvest is figured, ye Reformation of ye last century. The
Vintage is now to come: from anno 1689 or thereabout and
to last til the full destruction of ye Roman Babylon."[52] For
Evelyn the world would not end in 1689; like Cressener,
Lloyd, Jurieu, and possibly Burnet, he assumed rather that in
1689 or thereabouts the tide would turn against Babylon.
From the late seventeenth century onward, the reformed
church would assume the ascendancy in Europe and the task
of its leaders would be to convert the recalcitrant to the Protestant way and thereby ensure their reign in the millennial
paradise. After many misgivings and uncertainties, Evelyn and
his fellow churchmen came to see the Revolution as a step
along the prophetic path revealed in the Scripture. By 1692,

[50] Evelyn to Sancroft, Oct. 10, 1688, Evelyn MS 39, ff. 104–105,
Christ Church.
[51] See pp. 85–87.
[52] Notes after Revelation XXII, Evelyn MS 46, Christ Church.

Tillotson recorded in his private commonplace book, "I look at the King and Queen as two angels in human shape sent down to us to pluck a whole nation out of Sodom that we may not be destroyed."[53] A year earlier Thomas Burnet rejoiced to John Patrick, brother of Simon, about English victories in Ireland, worried about the military campaign on the Continent, but concluded optimistically: "The Resurrection of the Witnesses goes on very well in Savoy and Dauphiné, and that another argument to hope all will end well."[54] In 1694, Drue Cressener was still pursuing his study of Scripture, and he assured Simon Patrick that only in the reign of Justinian had the Roman church ever been completely in control of state affairs.[55]

If we need any further proof of the respectability of millenarianism within Anglican circles we need only briefly discuss the career of William Lloyd (1627–1717). Amid the various prophetic commentators so far discussed he probably inspired the greatest confidence as an able and learned interpreter. As bishop of St. Asaph and later of Worcester, he conferred widely with his devotees in the church and at court. During the Restoration he had been a thoroughgoing royalist and had prospered; after 1688 he was once again the object of considerable preferment.[56] His close friends included Thomas Tenison, John Evelyn, and Henry Dodwell, whose nonjuring after the Revolution led to a lessening of contact between the two scholars.

Lloyd became one of the most prominent latitudinarians in the post-Revolution church. He was an original drafter of the abortive Comprehension Bill, an effort with which he had

[53] Quoted in D. D. Brown, "An Edition of Selected Sermons of John Tillotson (1630–1694) from MS Rawlinson E. 125 in the Bodleian Library . . ." (M.A. thesis, University of London, 1956), lxi.

[54] Tanner MS XXVI, f. 44, Bodleian Library.

[55] Tanner MS XXV, f. 216, Bodleian Library.

[56] G. V. Bennett, "William III and the Episcopate," in G. V. Bennett and J. D. Walsh, eds., *Essays in Modern English Church History* (London, 1966), 128.

been concerned for many years.[57] As bishop of St. Asaph during the Restoration, Lloyd had made acquaintance with such prominent Dissenters as William Penn.[58] The concern of Lloyd, as of many other low-churchmen, was to comprehend the moderate Dissenters within the church and thereby to strengthen its hegemony at the expense of the divisive Protestant sects. In the 1695 commission set up by William III to regulate ecclesiastical preferment, Lloyd was a member along with Stillingfleet, Sharp, Tenison, Patrick, and Gilbert Burnet. Lloyd's single-minded devotion to the church and his willingness to propagandize the righteousness of the 1688–1689 settlement were political activities entwined with his deeply held millenarianism.

Lloyd's influential position within the ecclesiastical establishment gives added importance to his role as an almost official interpreter of the Scriptural prophecies. In June 1688, Lloyd announced on the basis of his calculations that popery could not last another year—presumably in England.[59] Spurred on by the Revolution, Lloyd predicted in 1691 that the Ottoman Empire would be rendered harmless and that the kingdom of France would be destroyed.[60] All of these predictions were based on Lloyd's interpretations of the cryptic prophecies in the books of Daniel and St. John, and he willingly shared his conclusions with his church colleagues. In 1712 he assured Queen Anne that in four years the church of Rome would be destroyed and the papal city consumed.[61] In March 1703/4, Lloyd told David Gregory that the possible date for the destruction of antichrist was 1736. When Gregory observed that Joseph Mede's date of 1715 was more generally accepted,

[57] A. Tindal Hart, *William Lloyd, 1627–1717* (London, 1952), 134. Cf. Record of a public debate between Lloyd and the Presbyterians, MS 91.17, Dr. Williams's Library. Lloyd would not accept that presbyters and bishops were in any way interchangeable.
[58] Ibid., 41–42.
[59] T. Birch, *The Life of . . . John Tillotson* (London, 1752), 168,
[60] Luttrell, II, 213. Cf. Hart, 137.
[61] Hart, 177.

Lloyd responded that he was convinced his own calculations were correct, but hoped that Mede's more optimistic date would prove after all to be true.[62]

Prophetic accuracy was at best precarious, and Lloyd always admitted the possibility that his calculations could be off by a few years. When the conditions of the French Protestants worsened—their triumph was an especially important stage in the fulfillment of the prophecies—Lloyd confided to Tenison: "They [the Vaudois] are indeed now in ye hands of their bloody enemies, and nothing can preserve them from destruction but some extraordinary appearance of ye hand of God. And what that can be, I cannot guess. But if they are destroyed, it breaks my whole scheme of Interpretation of ye Prophecies of Scripture. Let that be as it please God."[63] Lloyd seldom, however, abandoned hope. Four years after his despairing letter to Tenison he once again noted renewed strength among the Vaudois.[64]

Given Lloyd's deep commitment to millenarianism, his reluctance to make his prophecies and predictions completely public causes some bewilderment. Yet the delays in publication experienced by Cressener and Burnet in mid-1688 should provide a clue to explaining the public shyness of Anglican millenarians. At the time of Cressener's and Burnet's difficulties, Lloyd's chronology, which was also in the press, disturbed, if not offended, certain unknown members of the church.[65] A few years later other millenarian speculators in

[62] W. G. Hiscock, *David Gregory, Isaac Newton and Their Circle* (Oxford, 1937), 16. For continuing interest in the millennium in sectarian circles see Robert Prudom, *The New World Discovered, in the Prospect Glass of the Holy Scripture: Being a Brief Essay to the Opening Scripture Prophecies, Concerning the Latter Days* (London, 1704).

[63] MS 930, f. 42, Oct. 17, 1696, Lambeth Palace Library.

[64] MS 1029, f. 109, Dec. 27, 1700, Lambeth Palace Library.

[65] William Lloyd, *An Exposition of the Prophecy of Seventy Weeks, Which God Sent to David by the Angel Gabriel* (London, 1690?); Charles Hatton to Viscount Hatton, June 18, 1688, MSS ADD Hatton-Finch 29573, f. 167, B.L.

the church, for example Lloyd's good friend Pierre Allix, aroused hostility within and without church circles.[66] The most plausible explanation for this hostility takes us back to the dangerous association of millenarianism with radical politics. When Lloyd was finally attacked publicly in 1710 by a cleric in his own diocese, he was accused of lending support to the French prophets, extreme enthusiasts who from an Anglican perspective challenged the authority of church and state. The charge so deeply disturbed Lloyd that he was at pains to remind John Sharp, archbishop of York, that he had never predicted the fall of the French king, merely the fall of the French empire.[67] Given the antimonarchical nature of radical millenarianism, it behooved Lloyd to make that distinction absolutely clear.

Yet the court appears never to have suspected Lloyd of republican sentiments. Queen Mary encouraged his millenarian speculations. In 1694, Lloyd wrote to Stillingfleet, "Her Majesty does me great honour in taking notice of my Chronology and I shall return to it the sooner because she is pleased to call for it."[68] Such royal encouragement did not cease with Mary's death. Lloyd was a frequent visitor at the court of Queen Anne, and John Sharp, archbishop of York and confidant to the queen, urged Lloyd to continue with his attempt to apply prophecies to modern events "which may [be] of no great moment towards the conversion of the Jews." Sharp assured him that the queen herself was very keen for Lloyd to finish his study and present his findings.[69]

Other millenarian divines such as the Presbyterian, Robert Fleming (1660?–1716), found favor with William III. He consulted Fleming on ecclesiastical affairs in Scotland and al-

[66] Lloyd to Allix, July 12, 1697, MS 953, f. 54, Lambeth Palace Library.

[67] Sharp MSS, Borthwick Institute microfilm, 3/0. Cf. Hart, 199–201. Hart mentions Laughton's attack, but does not give the reason for it.

[68] MS 201.38, f. 69, Dr. Williams's Library.

[69] Sharp MSS 3/S, 1704.

though we know nothing about William's reaction to Fleming's millenarianism, certainly his predictions could not have given offense. In 1706, Fleming preached that the Revolution of 1688–1689 had been a stage in the providential plan leading to Protestant victory in Europe and the fall of the French monarchy before 1794. His prophecies were reprinted during the French Revolution.[70]

At a time of national crisis produced by the Continental wars Anglican pulpits, too, resounded with attacks on the beast and predictions of its defeat. After the Battle of Blenheim that staunch defender of the latitudinarians during the Restoration, Edward Fowler, now bishop of Gloucester, described the victory as a "pledge of an infinitely more glorious one in due time to come, *viz.* the perfect salvation the Messiah should bring, at His Second Coming, to all His people throughout the World."[71] Fowler was deeply committed to "the doctrine of the Millenium; or The most happy and pacifick State of the Church in future times" which had been ignored for many centuries until its "happy resurrection since the Reformation; and no man was more instrumental to it than our most Learned Mr. Joseph Mede."[72] Fowler's millenarianism, in defense of war and aggressive foreign policy, was undoubtedly sincere, but prophecy in the hands of the less scrupulous propagandists of the time, such as Defoe, was also used simply to entice the credulous into supporting the government's actions.[73]

Perhaps that usage coupled with the radical associations of millenarian prophecy contributed to the decline of such spec-

[70] Fleming, *Seculum Davidicum Redivivum: The Divine Right of the Revolution Scripturally and Rationally Evinced and Applied* (London, 1793). Delivered as a sermon June 27, 1706.

[71] Fowler, *A Sermon Preached in the Chappel at Guildhall, upon . . . 7th of September, 1704 . . .* (London, 1704), 2.

[72] E. Fowler, *Reflections upon the Late Examination of the Discourse of the Descent of the Man-Christ Jesus from Heaven Together with His Ascension to Heaven Again* (London, 1706), 114–115.

[73] See Rodney M. Baine, *Daniel Defoe and the Supernatural* (Athens, Ga., 1968), chap. 5.

ulations within church circles. The generation of older latitudinarian divines that governed the church from 1689 until
their deaths generally before 1720 provide in their letters and
diaries, and only occasionally in their sermons, ample evidence that millenarianism flourished among them. Well into the
reign of Anne, millenarian speculators such as William Lloyd,
or his friend Pierre Allix, were making predictions. Allix received patronage from Gilbert Burnet and John Tillotson as
well as from Lloyd, and his manuscripts in the Cambridge
University Library reveal that in 1710 he was busily at work
on the meaning of the prophecies and that he pondered the
work of Mede and other commentators with absolute seriousness.[74] After 1714 millenarianism appears to be in complete
decline in church circles although it can readily be found
even later in sectarian Protestantism. John Wesley set the date
1836 for the probable destruction of antichrist and the coming
of the new heaven and the new earth.[75]

Certainly the radical implications of millenarian speculations made their public revelation dangerous to the maintenance of social and political stability as churchmen conceived
it. Doctrines that cannot be enunciated frequently from the
pulpit are in danger of simply being forgotten by the faithful.
Yet intellectual elites often maintained esoteric doctrines that
survived undiluted from one generation to the next. The explanation that millenarianism declined in the church because it
was tainted by radicalism or was unsuited for sermonizing
does not appear to be fully adequate. The millenarians and
their friends and supporters knew perfectly well that Anglican

[74] Great Britain, *Calendar of State Papers, Domestic Series, William
and Mary, 1, 1689–90* (London, 1895), 245; MSS ADD 2608, U.L.C.
Cf. *Proceedings of the Huguenot Society of London,* 13 (1923–1929),
625–627, 248–254. For further evidence of millenarianism in church
circles see Cotes to John Smith, 1703, Turner-Dawson Collection,
R.4.42, Trinity College, Cambridge.
[75] LeRoy E. Froom, *The Prophetic Faith of Our Fathers. The Historical Development of Prophetic Interpretation* (Washington, D.C.,
1948), II, chap. 26.

millenarianism, properly understood, offered no solace to the radicals; indeed it took one of their most fertile doctrines and interpreted it in such a way as to undermine radical politics. The explanation for the decline of millenarianism must be found elsewhere, and it is tempting to see a relationship between the decline of speculations so manifestly "unscientific" and the rise of Newtonian science.

The temptation fails, however, when we realize that the followers and associates of Newton and his natural philosophy, as well as the master himself, belonged, by and large, to the millenarian circle within the church we have just described. Thomas Burnet and Newton were on good terms in the 1680s, and they corresponded about Burnet's theories concerning the origin of the world. Newton expressed only minor reservations about those theories.[76] William Whiston and Richard Bentley attacked Burnet's theories, either because they were based on the mechanical principles of Descartes or because, in John Keill's view, all "world-making" endangered piety.[77] Yet none of these divines repudiated either Burnet's explicit or implicit millenarianism.

William Whiston belonged to the Newtonian clique during the 1690s and throughout most of the reign of Anne,[78] finding himself excluded only after he persisted in publishing his anti-trinitarian views. In 1707 he gave the prestigious Boyle lectures and publicly affirmed his faith in the accomplishment of the Scriptural prophecies. The sponsors of the lectureship,

[76] H. W. Turnbull, ed., *The Correspondence of Isaac Newton*, II (Cambridge, 1960), esp. 329–334.

[77] Discussed in a long essay, privately communicated, by David Kubrin, "How Sir Isaac Newton Helped Restore Law 'n Order to the West" (Washington, D.C., 1972), pt. 2, 69–83; pt. 1 published in *Liberation*, 16 (1972), 32–41. I have reservations about Kubrin's theory of Keill's influence on Newton, but I hope that his essay will soon see publication.

[78] For example, see Whiston to John Sharp on Bentley's assistance to Whiston, 1702, Box 3N, Sharp MSS, Borthwick Institute. By 1708, Lloyd was furious with Whiston for having been "carried away as you have been," see MSS ADD 24197, B.L.

Tenison and Sir Henry Ashurst being the most prominent at this time, could only have approved of its contents.[79] When Whiston lectured, the English church was under attack from the French prophets who were themselves millenarians,[80] and to add to the church's plight the cause of international Protestantism hung on the outcome of the Continental wars.

Whiston propagandized for English victory by elevating the war to the status of a holy war. He equates the French kingdom with the beast of the Old Testament and argues that by the destruction of this portion of the Roman Empire the millennium draws closer. Whiston explains that "a Beast in Prophetic Stile always denotes an empire . . . and it will be easily understood, that by those four great Beasts were denoted the four grand Monarchies, which God permitted to tyrannize over his Church for many Ages; the Babylonian, the Medo-Persian, the Grecian, and the Roman."[81] A victory over Louis XIV would be a victory of European Protestantism and would draw the Reformation to its fulfillment.

Whiston bases his understanding of the Scriptural texts upon a literal reading of their contents. With the assistance of this reading, and of the work of Joseph Mede, Whiston argues that the world will be destroyed by a universal deluge,[82] as distinct from Burnet's universal conflagration, and that upon its destruction the second coming of Christ will be on hand. This universal destruction of the world will fulfill God's promise to destroy his wicked people. Whiston states further that only the "very latest and best of all" discoveries in natural philosophy support his interpretation of the world's course.[83] Whiston is, of course, referring to the Newtonian system of

[79] Whiston dedicated his lectures to Thomas Tenison, archbishop of Canterbury, and to Sir Henry Ashurst, both trustees of the lectureship. Evelyn had died in 1706.
[80] See Chapter 7.
[81] Whiston, *The Accomplishment of the Scripture Prophecies* (London, 1708), 10.
[82] Ibid., 94–95.
[83] Ibid., 95.

which he was an important proponent and explicator. In Whiston's mind, Newtonian physical laws and the prophetic texts of Scripture complement and augment their mutual importance. That Whiston could see no contradiction between his science and millenarianism is yet another illustration of how crucially important for the church was Newton's reliance upon divine will as active in the universe. Churchmen, like Whiston, who justified the church's interests by recourse to the Scriptural prophecies could also enlist the physical order, as explained by Newton, to illustrate the power and efficacy of providence.

The order and regularity of the Newtonian universe, Whiston argued, will be maintained and reflected in the temporal order and security promised to a nation fulfilling God's commandments.[84] If a nation deviates from God's dictates, then physical and material disorder, famine, earthquakes, unholy war, and pestilence await it. As an alternative to this unhappy prospect, Whiston pleads for a stable polity, with Newton's laws of motion as his model. At the same time he plans optimistically for the millennium, for the destruction and transformation of that polity through physical means, a providential action which was totally in keeping with Newton's concept of God's active role in the physical order.

In the Boyle lectures Whiston is careful not to set a date or time for the accomplishment of the prophecies. Yet just a year before delivering his sermons Whiston proclaimed that the prophecies of Revelation were now being fulfilled, the beast was near destruction, and his demise would commence with the sounding of the seventh trumpet.[85] Being more explicit than any of his contemporaries in the church, with the exception of Lloyd who may have assisted in the writing of Whiston's *An Essay on the Revelation of Saint John* (1706), in that book Whiston proclaims 1716 as the date for the de-

---

[84] Ibid., 188–189.
[85] *An Essay on the Revelation of Saint John* (London, 1706), 99.

struction of the beast.[86] In unison with other millenarian churchmen Whiston envisaged no major readjustment of the existing political and social structure either in preparation for the imminent reign of the saints or during their millenarian ascendancy.

Once again the writings of Whiston, like those of Burnet, reveal a shift from the explicit millenarianism of *An Essay*, completed in late 1705 and published in 1706, to the implicit millenarianism and cautious tone of the Boyle lectures delivered in 1707 and published in 1708. The only possible explanation for this change appears to lie in the furore caused in 1706 by the enthusiastic French prophets and the danger to political and social stability churchmen perceived in their movement. Whiston toned down his millenarianism in the Boyle lectures, not because he believed less in what he had said previously but because it was politic to do so. In 1734 he still subscribed just as fervently to millenarianism, and he took issue with Newton, who had died in 1727, because his chronology "puts the End of the papal Dominion so many Centuries later that the End of its Supporters the ten Kings, and than all the collateral Prophecies will admit."[87]

Whiston's quarrel with Newton merely illustrates the differences of opinion among the church prophets.[88] Given the Scriptural texts with which their rich imaginative faculties worked, differences in interpretation were inevitable. Whiston's comments indicate that Newton did project a date for the destruction of the Roman Empire, one that Whiston felt to be too far in the future. Newtonian scholars, however, have

[86] Ibid., 270–272.
[87] Whiston, *Six Dissertations . . . Remarks on Sir Isaac Newton's Observations upon the Prophecies of Daniel* (London, 1734), 297–302. For insight into Whiston's piety and sensibility see his "Commonplace Book," MS ENG. MISC. d. 297, Bodleian. In 1748, Whiston believed the millennium to be but twenty years away; Henry Pettit, ed., *The Correspondence of Edward Young, 1683–1765* (Oxford, 1971), 302.
[88] For example, P. Allix, in *Two Treatises* (London, 1707), quarrels with Whiston.

either ignored his millenarianism or, when they have dealt
with his study of the prophecies, have laid emphasis on his
hesitation and cautious refusal to predict the future: "The
historical excluded the metaphysical as well as the mystical."[89]
Yet Newton may have thought that the world would end in
the year 2000;[90] Whiston, Lloyd, Burnet, Evelyn, probably
Cressener and Allix, would have disagreed. But publicly, and
as far as the present manuscript sources reveal privately, Newton eschewed prediction because he thought it precarious and
chancy. There is no evidence, however, that he regarded the
millenarianism of other churchmen who did make exact predictions as "mystical effusions."[91] Newton's caution does not
render his historical studies and speculations on the prophecies
qualitatively different from the millenarian beliefs and concomitant historical research and chronological calculation of
any other Anglican millenarian. There is no evidence of substantial intellectual disagreement even between Newton and
his intimate associate Nicolas Fatio de Duillier, one of the
eventual devotees of the French prophets. Their only disagreement centered on what Newton called "fansy." He
wrote cryptically to Fatio, "I am glad you have taken ye
prophecies into consideration & I believe there is much in
what you say about them, but I fear you indulge too much in
fansy in some things."[92]

Newton's caution was merely a more extreme version of

[89] Frank Manuel, *Isaac Newton Historian* (Cambridge, 1963), 146–
147, 156.

[90] "Prophecy of Seals and Trumpets Synchronized with Pouring out
of Seven Vials and Seven Last Plagues" (1691), MS Locke C.27, f. 88,
Bodleian; but no dates are given. William H. Austin, "Isaac Newton
on Science and Religion," *Journal of the History of Ideas*, 31 (1970),
525. Cf. L. Trengove, "Newton's Theological Views," *Annals of Science*, 22 (1966), 277–294.

[91] The term is Manuel's, 155. He appears to have altered slightly his
views; see his *The Religion of Isaac Newton* (Oxford, 1974), 99–100.

[92] H. W. Turnbull, ed., III, 245, Feb. 14, 1692/3. For an interesting
hypothesis on their intellectual rapport see D. P. Walker, *The Ancient Theology* (London, 1972), 260–261.

that found among other churchmen, and in every case they restrained themselves in order to serve the church's interests and not because they regarded predictions as less scientific or more mystical than had students of the prophecies in ages gone by.

"Why should we not think that the Prophecies wch concern the latter times into which we are fallen were in like manner intended for our use that in the midst of Apostacies we might be able to discern the truth and be established in the faith thereof?"[93] Newton answered his own rhetorical question: the prophecies were given for the church so that by searching them it would recognize Christ and not succumb "in this desperate age" to the deceptions of antichrist. The prophecies revealed to Newton the ever-present, ever-operative providence of God at work in history. "God had revealed himself to Newton not only in the order of nature, which he had interpreted mathematically in his studies of philosophy, but also in myth and prophecy. All were traces of one Creator."[94]

Newton's firm and essential belief in divine participation and regulation of the world natural and the "world politick" may explain the occasions when he envisaged disorder and even chaos in the natural order. In the first edition of the *Principia* (1687), Newton implied that he no longer accepted, as once he had, the continuous and harmonious interaction of the various parts of the cosmos. The earth could increase in size until the balance of the universe had been upset.[95] Throughout the 1690s reports reached the Royal Society about Newton's elaborations on this notion. In 1697, Halley read a paper to the Society wherein he described the effects "of a Collision of a great Body such as a Comet agst the Globe of

---

[93] "A Treatise on the Book of Revelation," a Newton manuscript, Yahuda MS 1, ff. 2–3, Jewish National and University Library. Available on microfilm, no. 664, U.L.C. Published in part in Manuel, *The Religion*, Appendix A, 109.

[94] Manuel, *Isaac Newton Historian*, 164.

[95] Kubrin, "Newton," 337.

the Earth, which he described might be undisproved of times before Adam, wherein possibly that Earth might have been reduced to a Chaos, after having been for many years together such as now it is."[96] In this report Halley quite probably spoke for Newton; five years later David Gregory recorded: "The Comet whose Orbit Mr. Newton determines may sometime impinge on the earth. Origen relates the manner of destroying the world's by one falling on another."[97] When Newton contemplated the destruction of our world he may have been trying to establish natural principles for the apocalypse or he may simply have been expressing a pessimism, common in church circles, that saw destruction as an ever-present possibility for a sinful nation toying with obedience to the providential will. Both sentiments were sides of the same coin; both relied on certain unshakable assumptions.

Providence operates in history and in nature; all the traces of the creator are intended never simply to enhance personal knowledge and piety but to enlighten the whole church—specifically the English church—whose task is to complete the Reformation and prepare the world for the second coming, the new heaven and the new earth. Like his latitudinarian colleagues Newton searched for a broad Christianity to be embraced by the church as a Protestantism acceptable to all people.[98] In adopting latitudinarian natural religion the church would in the process purify itself, revitalize its doctrines, and arm itself for the struggle with antichrist.

In the light of Newton's involvement with the Boyle lectureship and its lectures[99] we can hardly doubt that he came to regard his natural philosophy as the foundation upon which such a latitudinarian natural religion would rest secure. Like

[96] MSS ADD 4478b, ff. 142–150, B.L.
[97] Gregory MS 247, f. 87, The Royal Society.
[98] Manuel, *Isaac Newton Historian*, 159. For the compatability of providence and natural philosophy, see J. E. McGuire, "Force, Active Principles and Newton's Invisible Realm," *Ambix*, 25 (1968), 154–208.
[99] See Chapter 4.

Isaac Barrow, Newton believed in the intimate relationship between the world political and the world natural; indeed his manuscript treatise, "The Language of Prophecies," provides a clear and unequivocal statement of the relationship.[100] His discoveries in natural philosophy explicated the workings of the world natural, and his natural philosophy with its absolute reliance on providence, its emphasis on design, order, and simplicity, when applied by his followers to the world political, was intended to synchronize these worlds and prepare them for the fulfillment of the millenarian prophecies.

As Newton first presented it to Bentley, and as he and his low-church friends understood it, Newton's science provided them with seemingly incontrovertible proof of God's providential design in nature as well as in the realm of human affairs. The natural religion developed by the latitudinarian preachers of the Restoration, based on the scientific knowledge then available, could now be secured on the firm foundation laid out by Newton in the *Principia* (1687). Latitudinarian natural religion and its inherent social ideology flourished with new vitality in the Boyle lectures, and the church faced the hostility of its political foes, the "crafty, ill-principled men," and the degeneracy of its age armed with the principles of Newton. The mathematical princples of natural philosophy, universal gravitation, the vacuum, the atomic structure of matter, and active principles were all weapons with which the destruction of antichrist would be ensured. The reasonableness of Christianity, or more precisely of latitudinarian natural religion, would secure the universal acceptance of Protestantism and its guardian, the English church.

Yet some of the most important Newtonians like Richard Bentley and Samuel Clarke were hardly millenarians. Bentley did believe in the final destruction of the world and preferred the mechanism of a deluge, rather than a universal conflagration, for its destruction. Only through the intervention of God

[100] H. McLachlan, ed., *Sir Isaac Newton Theological Manuscripts* (Liverpool, 1950). The epigraph for this book is from p. 120.

could the constant decay and final destruction of the world be prevented.[101] Bentley leaves us no historical evidence that I know of, however, to indicate that he urgently anticipated a millennial paradise to be established at the end of time. In the case of Clarke, Whiston claimed that he concurred with his prophetic musings as well as with those of Joseph Mede.[102] This is highly probable given Clarke's intimate association with Newton and the latter's deep interest in the prophecies. However, Clarke himself never seems to have ventured into millenarian speculation. He feared the enthusiasts of his time, but more important, his own work on the prophecies extends only to their application to Christology, and he constantly cautions that the difficult and obscure prophecies should be studied only by individuals skilled in history.[103]

Both Bentley and Clarke represent a young and rising generation of latitudinarians separated from Lloyd, Evelyn, Tenison, Newton, and the others by age and experience or the lack of it. They had been told about the civil wars and Interregnum; they had not experienced those events. Even in 1688–1689 they were just beginning their ecclesiastical rise, and the Revolution benefited them. For Bentley the Revolution hardly presented a moral issue; it had been a matter of self-interest.[104] Newton appears to have regarded Bentley as glib on the matter of the prophecies and on religion in general.[105] Al-

[101] A. Dyce, ed., *The Works of Richard Bentley* (London, 1838), III, 87.

[102] W. Whiston, *Historical Memoirs of the Life of Samuel Clarke . . .* (London, 1730), 156.

[103] Loose page dated May 19, 1719, Wake MSS, CCL, XVII, Christ Church Library; *A Discourse Concerning the Connexion of the Prophecies in the Old Testament, and the Application of Them to Christ* (London, 1725), esp. 39. At times Newton would have concurred; McLachlan, ed., *Sir Isaac Newton Theological Manuscripts*, 34.

[104] Dyce, ed., 228.

[105] Manuel, *Isaac Newton Historian*, 141, 288. Cf. William Whiston, *Memoirs of the Life and Writings of William Whiston . . .* (London, 1749), I, 106–109.

though close to their master, Bentley and Clarke possessed a different religious sensibility, that of a younger generation of latitudinarians, and it infused their sermons for the Boyle lectureship and hence the first public presentation of latitudinarian natural religion based on Newtonian principles.

The decline of millenarian speculation in church circles during the early eighteenth century raises the possibility that eighteenth-century ideas of progress may have arisen from a transformation of millenarian sentiments into theories about the perfectability of this world. Certainly Anglican preachers of the early eighteenth century can be cited who believed in a very Christianized concept of progress.[106] Indeed E. Tuveson has argued that in the writings of Joseph Mede, Thomas Burnet, and William Whiston one can already see the beginnings of this transformation. To him these latitudinarians "emphasize less and less the catastrophic aspect of the millennium, [and] more and more . . . draw it into the stream of history."[107]

Yet the research here presented into Anglican millennarianism does not, I believe, support Tuveson's interpretation. There is little sense of progress in the millenarianism of Burnet, Evelyn, and Whiston. True millenarianism did not provide the intellectual background to ideas of progress. Millenarianism was rather, and more simply, a common psychological response (which could be either optimistic or pessimistic) to political instability in seventeenth-century England. With the growth of political stability after the Revolution of 1688–1689 millenarianism gradually declined in establishment circles. Ideas of progress in Anglican circles during the eighteenth century owe their origin more to smugness than to a rethinking of the meaning of the millennium. The idea of progress also has its origin in the highly sophisticated sense of history found in the writings of early eighteenth-century

[106] R. Crane, "Anglican Apologetics and the Idea of Progress," *Modern Philology*, 31 (1934), 273–301, 349–382.
[107] Tuveson, 134.

deists and freethinkers. While churchmen were busily trying to unravel the meaning of Revelation, Toland and others were treating the Bible simply as a historical, human document.

Often in direct opposition to these freethinkers, latitudinarians like Bentley and Clarke preached their version of natural religion built upon the harmonious relationships they imagined between the "world politick" and the world natural. They intended this natural religion to serve the immediate needs and interests of godly men and to thwart the mounting attack being launched by the deists, freethinkers, and atheists. The Newtonians realistically perceived that the church had been weakened by the Revolution and proposed their natural religion as the only bulwark against the ravages of self-interest and atheism. If they imagined any sort of utopia it would have been a state where stability and order predominated, where the principles of Newton ruled nature and presented a model for the ordering of society. The saints in such an order would be godly and industrious men who drew inspiration from the providential stability of the Newtonian universe and who curbed their self-interest to serve the necessities of stability and prosperity.

Political and social stability returned to England in the decades after 1688–1689, and with the growth of English power the threat posed by antichrist, embodied in the French empire, lessened. This factor had far more to do with the virtual disappearance of millenarianism in church circles than did the rise of science. Indeed millenarianism died with the older generation of latitudinarians whose religious sensibility had been forged by the seventeenth-century revolution. In the face of that upheaval they had turned to science, political and ecclesiastical moderation, and reason to counter the threat posed by the radicals and to win back into the church what Barrow called men of "business and dispatch." For the churchmen of the Restoration natural religion served as a means to the end: the establishment of a millenarian paradise where finally the church would attain the supremacy denied it by the Revolution.

Newton belonged with the older generation of low-churchmen whose sensibilities had been formed by the mid-century Revolution; his followers who gave the Boyle lectures by and large did not. In their hands the Newtonian natural philosophy found application to the dramatic change in the church's fortunes occasioned by the Revolution of 1688–1689. The main issue now was not to lead the nation in preparation for the second coming; rather it was to hold on to what remained of the church's political and moral power and to use it as best they could. The Revolution Settlement had established Protestant pluralism, doomed comprehension, and seemingly opened the Pandora's box of freethinkers, deists, Socinians, atheists, and other equally undesirable nonbelievers. The preaching of natural religion based on solid and irrefutable natural philosophy assumed critical importance; the only form of Protestant unification possible seemed to rely on the widespread acceptance of a liberal Christianity, not on an amalgamation of the churches. The young Newtonians grasped the issues facing the church and addressed themselves to finding an intellectual formula that would win acceptance.

The scientific principles upon which their formula came to rest and the podium from which they preached were given to them by their intellectual leaders and patrons—Boyle, Newton, Evelyn, and Tenison, among others. For them millenarianism gave meaning to political and natural events, and they applied to the understanding of those events God's word as revealed in Scripture and God's work as revealed by science. In the cases of Burnet, Whiston, and probably Newton, millenarianism even fostered their scientific enquiry in that, because they accepted the prophecies, they sought to understand the means chosen by providence to control the natural order and finally to authorize its destruction.[108] Furthermore, mil-

[108] Cf. H. R. Trevor-Roper, *Religion, the Reformation and Social Change* (London, 1967), 47.

lenarianism made urgent the necessity of finding a natural religion acceptable to all Protestants. Without the special ingredient of millenarian fervor—if that is the way to describe the secretive comings and goings and private conferences of our churchmen—the impulse to proclaim a natural religion based on the operations of nature would have been more tenuous, less certain of itself. Millenarianism aided in the acceptance of scientific enquiry and served to focus the application of natural philosophy to social and political issues. The natural religion of the latitudinarians, the ease with which they embraced the new science, constituted, as their critics were quick to note, a new and dangerous version of Protestantism wherein reason dominated both revelation and faith. And in large measure their critics were right.[109] Latitudinarian natural religion, especially as explicated by the Boyle lectures, deals with religion almost solely as a device for curbing self-interest and maintaining social stability—all in imitation of the Newtonian model of the universe. In the late seventeenth century natural religion accommodated itself to the needs of a market society. Many of its advocates, however, assumed that they were slaying the antichrist, not jousting with Mammon.

[109] Cf. John Dillenberger, *Protestant Thought and Natural Science* (London, 1961), 109.

CHAPTER 4

The Church, Newton, and the Founding of the Boyle Lectureship

The preaching and writings of Restoration liberal church-men, reinforced by their private millenarianism, provided a firm intellectual foundation upon which the second generation of low-churchmen built during the 1690s and beyond. The careers of these younger divines, such as Richard Bentley, Samuel Clarke, and William Whiston, received assistance and encouragement from Stillingfleet, Moore, Evelyn, Tenison, and Newton, among others. As we shall see, they provided these younger colleagues with a special and prestigious podium, the Boyle lectureship, from which to reach the wealthiest and most influential congregations in London.

After the Revolution of 1688–1689 the church was secure, yet endangered. This paradoxical situation is reflected in lati-tudinarian behavior. On one hand, Anglicans of liberal per-suasion were dominant in the intellectual and ecclesiastical life of the nation. Yet their sermons reveal unease over the correct-ness of the political settlement and, more noticeably, an urgent desire to secure the church's position of moral and social leadership. The Revolution demanded from churchmen a re-definition of the role of providence in political affairs, and it also required a reinterpretation of the church's mission. In consequence of the Toleration Act and the new authority accorded to Parliament, only a broad natural theology, anchored on sound natural philosophy, could enable the church to propose an acceptable Protestantism suited to an open society wherein rival religious, political, and social groups vied for their share of profits and preferment and therefore for their own power and self-interest.

The Revolution secured Anglican hegemony, yet weakened the church's political power. The latitudinarians realized that their natural religion, if properly articulated, could forge among Protestants a consensus upon which the church's moral leadership would rest secure. The Revolution Settlement must not be allowed to become a victory for the "crafty and ill-principled"; Hobbism must not be confirmed as triumphant. All of those sentiments can be found in the will of Robert Boyle, who in July 1691, a few months before his death, established the Boyle lectureship. The lectures would be "for proving the Christian Religion, against notorious Infidels, *viz.* Atheists, Theists, Pagans, Jews, and Mahometans, not descending lower to any Controversies, that are among Christians themselves: These Lectures to be on the first Monday of the respective months [excluding June, July, August, and December] . . . and to answer such new Objections and Difficulties, as may be started, to which good Answers have not yet been made."[1]

As he instructed them, Boyle's friends established his lectureship and through it the younger generation of latitudinarians proposed Newton's system of the world as the "good answer" to "new objections and difficulties." Since the Boyle lectures first articulated one of the dominant versions of eighteenth-century Newtonianism, we should explore how and why the lectureship was created, who controlled it, and what role Newton himself played in this first popular exposition of his philosophy.

Boyle's concern for the maintenance of a moderate Christianity had been of long standing. As a young man, the youngest son of the powerful earl of Cork, he had faced the trau-

---

[1] The will was first published in Eustace Budgell, *Memoirs of the Lives and Characters of the Illustrious Family of the Boyles* (London, 1737), Appendix, 25. It was reprinted in Thomas Birch, ed., *Works of the Honourable Robert Boyle* (London, 1744), I, 105.

matic dislocation presented to men of his class by the civil wars. He had come to terms with that dislocation and the concomitant intellectual revolution by a painstaking evaluation of his own ethical obligations and of the role played by religion in society. His ethical treatises from the 1640s, still extant in manuscript form, partially reveal the process Boyle underwent as he devised his unique accommodation to the often conflicting needs of piety and self-interest.[2]

His eventual accommodation had important results. Boyle became the most famous virtuoso and natural philosopher in seventeenth-century England. His religious and philosophical writings were a guide for an entire generation of religious thinkers, many of whom composed the moderate faction of the church.

After 1688–1689 it can be said that the latitudinarians had to come to terms with problems similar to those faced by Boyle in the 1640s. The dislocation of the church produced by the Revolution of 1688–1689 bears some analogy to the dislocation produced by the civil wars. To be sure, the similarity of these historical phenomena is only one of analogy. But the accommodations made by moderate Protestants at mid-century[3] prepared the way for the final accommodation of religion to the needs of an ordered, acquisitive society made by churchmen at the end of the century.

As had Boyle a few decades earlier, Richard Bentley, John Harris, Samuel Clarke, and William Derham turned to the model presented by the natural world in an effort to develop a new way of thinking about society and the role of God in its workings. The Boyle lectureship provided the opportunity for them to explain the results of their thinking about nature and about society, and in their lectures Bentley, Clarke, Harris,

[2] See J. R. Jacob, *Robert Boyle and the English Revolution* (New York, 1977).

[3] Occasionally John Tulloch takes this approach. See *Rational Theology and Christian Philosophy in England in the Seventeenth Century* (London, 1874), II, 439.

and Derham became the first popular commentators on the Newtonian natural philosophy. Without their lectures, the new Newtonian philosophy would not have existed by the early eighteenth century as a coherent system to be understood by anyone outside the rather small circle of Newton's scientifically trained followers. The cogently reasoned arguments of the Boyle lecturers presented a formidable structure in support of natural religion, that is, in support of the religious principles advanced by the moderate segment of the church after 1688.

Although this chapter is not concerned primarily with the contents of the lectures, it might be helpful to discuss their corpus and to indicate the reasons for selecting certain lectures as more important than others. From the list provided in the appendix, it is evident that the Boyle lectureship was dominated during its early years by churchmen who were also important followers of Sir Isaac Newton. This coincidence was hardly accidental. All of the lectures of this period concentrated on developing a viable natural religion. The arguments given by lecturers such as John Williams, Samuel Bradford, and Offspring Blackall, in support of the inherent truth of divine revelation, stressed the reasonable quality of revelation, the ability of man to affirm the existence of divine truth by relying solely on his own power of reason.

This argument was, however, no longer completely adequate. As we shall see in our discussion of the freethinkers, reason had many advocates, and freethinking could be made as reasonable as natural religion. As Boyle saw, a new approach was necessary to prove God's existence and his providential action in the universe. Such an approach was provided by the Newtonian commentators, and both at the time and in retrospect their lectures were outstanding. In his diary, Evelyn records his enthusiasm for Bentley's lectures,[4] and this enthu-

[4] E. deBeer, ed., *The Diary of John Evelyn* (Oxford, 1955), V, 94, 123.

siasm was shared by Thomas Tenison, trustee of the lectureship and after 1695 archbishop of Canterbury. Because Bentley delayed publication of his second series of lectures, he incurred for a time the archbishop's impatient disfavor.[5] Bentley had synthesized many elements in the thinking of the moderate church, and his synthesis was repeated, clarified, and made more rigorous by the Newtonian commentators who succeeded him in the lectureship. In providing for the lectureship, Boyle had once again served well the interests of the church.

The codicil to Boyle's will establishing the lectureship was registered on July 28, 1691.[6] The codicils that dealt with Boyle's concern for religion were known in July 1691 even to churchmen outside the immediate circle of his friends. In his manuscript journal book, White Kennett recorded Boyle's intentions, revealed in the codicil of July 18, to found a society for the propagation of Christianity in foreign lands.[7] Among those who also obviously knew of Boyle's intended lectureship were the four trustees named in the codicil to administer it. They were John Evelyn, Thomas Tenison, Sir Henry Ashurst, and Sir John Rotherham.[8] Each was chosen because of his long friendship with Boyle and, we suspect, for his involvement in ecclesiastical and social affairs. Evelyn was intimately associated with church and scientific circles, as well as being personally entrenched with certain aristocratic families. In 1689, Tenison was made archdeacon of London and as a result

[5] C. Wordsworth, ed., *The Correspondence of Richard Bentley* (London, 1842), I, 112-113. The lectures appear never to have been published.

[6] The original codicil is to be found at Chatsworth House, Derbyshire. Copies of Boyle's will and the codicils are in the hands of Messrs. Currey & Co., 21 Buckingham Gate, London, the present trustees of the Boyle estate.

[7] Lansdowne 1024, f. 77, B.L.

[8] Cf. R. E. W. Maddison, "Studies in the Life of Robert Boyle, F.R.S. Part III," *Notes and Records of the Royal Society of London,* 10 (1952), 15-27.

had most of the parishes of London under his jurisdiction.[9] Since the lecturer was to be chosen from the London area, Tenison was in the best position to ascertain the ability of any candidate. His own preferment advanced as the moderate party gained control over the post-Revolution church. By mid-1691 the moderates were in firm control with John Tillotson as archbishop of Canterbury. In 1692, Tenison was made bishop of Lincoln and in 1695 he succeeded Tillotson as archbishop. In the period between 1695 and 1701, Tenison was undoubtedly the most important leader of the church, although with the Tory reaction in the reign of Anne his power waned relative to that accorded to John Sharp, archbishop of York.

With one or two exceptions, the other trustees, Ashurst and Rotherham, played minor roles in the selection of the Boyle lecturers.[10] Thomas Tenison and John Evelyn's joint correspondence and Evelyn's diary for the 1690s reveal their dominating interest in the trust Boyle had bequeathed to them. The control they exercised would in itself suggest that the Boyle lectures were part of the program advanced by Anglican moderates after 1688. Other important evidence supports that interpretation of the purpose of the lectureship. In a manuscript letter deposited in the British Library and dated September 17, 1694, William Wotton wrote to John Evelyn about a letter Wotton had received from Richard Bentley. Bentley's letter, containing a request to be made of Evelyn, sought permission to change the date of one of his lectures from September 3 to the first Monday in December. Bentley had been detained in Worcester, Wotton explained, and this inconvenience coupled with Bentley's assumption that "ye design of ye Lecture would be better answered in December

[9] Edward Carpenter, *Thomas Tenison, Archbishop of Canterbury* (London, 1948), 120.

[10] John Evelyn, *The Diary and Correspondence* (Bohn ed., London, 1859), III, 376–377; deBeer, ed., V, 123, 160–161.

when ye Town would be very full, than in September when it is always thinner,"[11] led Bentley to desire Evelyn's consent to the change.

Luckily Evelyn drafted his extremely interesting reply to Wotton on a blank page left in his letter. This draft reply provides clear evidence that Heneage Finch, baron Guernsey and first earl of Aylesford, and William Wotton, known for his participation in the ancients-and-moderns controversy, were intimately involved with the Boyle lectures. Evelyn wrote: "The suddain Alarme we received from my daughter have, hurried us from Wotton sooner by some days, than was intended; hence I find I denyed me of the honor I should have received from Mr. Finch and yourselfe, for expectation of whose Returne to Albury, I had something to shew you from the Bp. of Lincoln [Tenison], that he was not unmindful of your Commands, or at all unsuccessful in his joining with my Wishes for one so worthy to continue what Mr. Bentley had so laudably begun."[12] Finch and his son's tutor, William Wotton, appear to have given certain "Commands" to Tenison concerning the choice of a proper successor to continue Bentley's work. His successor was John Williams, after the Revolution a chaplain to William and Mary and bishop of Chichester from 1696. His sermons defended natural religion and, in particular, explicated the notion of divine providence.[13]

[11] MSS ADD 28104, f. 18, B.L.

[12] Ibid., f. 19, Sept. 20, 1694. In a previous essay I mistakenly identified the "Mr. Finch" of this letter with Daniel Finch, brother of Heneage and second earl of Nottingham. After consultation with Henry Horwitz, who kindly offered the identification as that of Heneage, I came to concur with his view. For my purposes, either Finch will do since in this period they shared similar political views and were leaders of the church party in Parliament. For Daniel Finch see H. Horwitz, *Revolution Politicks* (Cambridge, 1968), and M. C. Jacob, "The Church and the Formulation of the Newtonian Worldview," *Journal of European Studies*, 1 (1971), 139. Evelyn visited on other occasions with Wotton and Finch, see deBeer, ed., V, 184n, 219.

[13] See Gerald Straka, *The Anglican Reaction to the Revolution of 1688* (Madison, Wis., 1962), 65–99.

Finch's involvement in the Boyle lectures should come as no surprise. He and his brother Daniel, second earl of Nottingham, led the church party in the 1690s, and winning the church's interests in Parliament meant little if simultaneously the church's moral and intellectual hold was being undermined. Piety, coupled with an interest in natural religion and the new philosophy, had long prevailed in the Finch household. The uncle of Heneage and Daniel Finch, Sir John Finch, discussed these matters with the learned Dr. Thomas Baines who also tutored Daniel for a time.[14] Sir John Finch has left behind a long treatise on natural religion, and the Nottingham Papers in the Leicestershire Record Office include Finch's commonplace books with extensive notations on the new science and natural religion.[15] If we want to know more about the diffusion of the new philosophy and natural religion in the 1650s and 1660s these papers would obviously merit study. In the 1690s when Heneage Finch concerned himself with the Boyle lectures and with the promotion of Wotton and Bentley in the church,[16] he was continuing a long family tradition of church patronage and commitment to latitudinarian religion. His brother Daniel similarly fostered the careers of key moderates until they reached positions in the hierarchy, and he too possessed a keen sense of who were the church's intellectual enemies.[17]

After Bentley and Clarke, Wotton was the church's most important young intellectual leader and propagandist. In June 1691 he transmitted to Bentley John Craig's instructions on understanding Newton's *Principia*.[18] This transaction occurred

[14] Horwitz, *Revolution Politicks*, 3–5.
[15] Boxes 4976, 4977, 4978, Leicestershire Record Office. See H. Horwitz, "The Work of Sir John Finch," *Notes and Queries*, 213 (1968), 103–104.
[16] Tanner MSS, XXV, f. 399, Bodleian.
[17] A. Tindal Hart, *The Life and Times of John Sharp, Archbishop of York* (London, 1949), 332.
[18] H. W. Turnbull, ed., *The Correspondence of Sir Isaac Newton*, III (Cambridge, 1961), 150–152. Cf. Henry Guerlac and M. C. Jacob,

about one month before the registration in late July 1691 of Boyle's codicil setting up his lectureship. Wotton's subsequent interest in the lectures, as revealed by Evelyn's draft letter of 1694, allows for the possibility that in 1691, when he transmitted Craig's instructions to Bentley, Wotton may have been involved in or have known about whatever discussions took place between Boyle and his intended trustees of the lectureship, Evelyn and Tenison.

Speculation about the extent of moderate Anglican involvement in the foundation and administration of the Boyle lectureship raises a series of interesting and important questions about the lectures. Why was Richard Bentley chosen as the first lecturer? What was Newton's role in the setting up of these lectures that relied so heavily on his thinking about the structure of the universe and the relationship between creation and its creator? These two questions are related because Bentley consulted with Newton directly in July 1691 and then again toward the end of his first set of lectures.[19] Later in the 1690s, Bentley was a member of an intimate circle that centered around Sir Isaac Newton.[20]

Could it be possible that Newton suggested to the trustees that Bentley become the first Boyle lecturer? The evidence for Newton's direct involvement with the Boyle lectures is highly suggestive, yet not conclusive.

Edleston's, and recently Turnbull's, dating of July 1691 for Newton's directive to the young Bentley, setting forth the preliminary steps necessary to understand the *Principia*, is extremely important.[21] Bentley later put his knowledge to use in the final Boyle lectures of 1692. In late July of 1691, Boyle registered the codicil to his will, and there is no reason to

---

"Bentley, Newton and Providence (The Boyle Lectures Once More)," *Journal of the History of Ideas*, 30 (1969), 307–318.

[19] Guerlac and Jacob, 313–314.

[20] James Monk, *The Life of Richard Bentley* (London, 1833), I, 96.

[21] Turnbull, ed., III, 155–156. See also Joseph Edleston, ed., *Correspondence of Sir Isaac Newton and Professor Cotes* (London, 1850), 273–275.

suppose that Boyle had kept his intentions secret until July
28, the actual date of registration. Certainly the codicil would
not have come as a surprise to the trustees named in it. Through
mutual interest and membership in the Royal Society, Evelyn's
association with Newton in this period may be taken as a fact.
Evelyn's and Tenison's intimate association with Stillingfleet,
to whom Bentley was chaplain, is certain. There is every
reason, therefore, to assume that when Newton was instructing
Bentley, Boyle's intentions were known to Evelyn and Tenison, as well as to the coterie of newly appointed low-church
hierarchy. One early nineteenth-century biography of Bentley
claimed that Stillingfleet then suggested that either Bentley or
William Lloyd should receive the first Boyle lectureship.[22]

The claim is a plausible one, yet in 1691, Evelyn was already
acquainted with the young Bentley. Evidence found in the
manuscript letters of Evelyn deposited at Christ Church,
Oxford, reveals that as early as 1686, Bentley was in communication with Evelyn, whom he addressed as "My Noble
friend."[23] In this earliest known Bentley letter, part of which
is difficult to decipher, he discusses literary and political news
from the Continent, sends a message to Glanvill about the
circulation of Fontenelle's *Pluralité des Mondes*, and assures
Evelyn of his gratitude. Evelyn and Bentley appear engaged
in lively correspondence on literary and philosophical matters.
Bentley informs Evelyn, "I have a book for you to subscrip
[subscribe] to that you will be kinder to than you were to
Milton," and he sends his regards to Evelyn's wife.

From the contents of this Bentley letter, it appears that
Evelyn probably did not need the suggestions of others when
he looked for a bright, young churchman to launch the Boyle
lectureship. Bentley already was associated with the low church,
and his chaplaincy to Stillingfleet, his friendship with William

[22] Anon., *The Life of Dr. Richard Bentley, 1662–1742* (London?,
1825?), 4. The British Museum copy belonged to Samuel T. Cole-
ridge.
[23] Evelyn Letters 155, Christ Church.

Wotton and Evelyn, among others, as well as his extraordinary skill as a classicist, made him an obvious choice for the important Boyle lectures.

These facts do not explain, however, why Bentley became interested in the Newtonian natural philosophy in the summer of 1691. It would be possible to argue that he was urged to his study by Wotton, or maybe by Evelyn, who as soon as they heard of the lectureship thought of Bentley as the candidate to be prepared for the task. Unfortunately there is as yet no evidence to support this conjecture.

On the basis of the available evidence, we cannot even be certain that as early as the summer of 1691 Bentley was being seriously considered by the trustees of the lectureship. He was not formally chosen until February 13, 1692.[24] Nonetheless, we know that in mid-1691 Bentley had directions from Newton to assist him with the geometry and astronomy in the *Principia*. At just this time Bentley was involved in the preparation of his famous critical edition of Manilius' *Astronomicon*, a work which he did not complete at that time, and which was published posthumously.[25] It is reasonable to conjecture that Bentley, faced with an astronomical and astrological text, turned for assistance to the most prominent mathematician and astronomer of the day and asked to be instructed on the most recent ideas available on celestial matters. It is possible that Bentley remembered Newton because of his lectures delivered when Bentley was a student at Cambridge.[26]

Whatever factors conspired to bring Bentley to Newton's attention, their acquaintance was of considerable importance, for Newton, as well as Evelyn and Tenison, played a part in

[24] deBeer, ed., V, 88–89. The apparent formality with which Evelyn refers to Bentley should not be taken for a lack of familiarity. Evelyn is a master of understatement.

[25] Monk, I, 48–51. See especially *M. Manilii Astronomicon ex recensione et cum notis Richardi Bentleii* (London, 1739), 64. I owe this point to Henry Guerlac.

[26] Ibid., 8. These lectures, if they were given at all, dealt only with mathematics and optics.

the formation of the first Boyle lectures. In a memorandum dated December 28, 1691, David Gregory records: "In Mr. Newtons opinion a good design of a publick speech (and which may serve well at ane Act) may be to shew that the most simple laws of nature are observed in the structure of a great part of the Universe, that the philosophy ought ther to begin, and that Cosmical Qualities are as much easier as they are more Universall than particular ones, and the general contrivance simpler than that of Animals plants &c."[27] This memorandum is significantly dated two days before Boyle's death. At that time, Boyle was in his last illness, and we know that he was concerned with his will and its proper execution. On December 29 he added a final codicil and appointed Sir Henry Ashurst to succeed the then deceased Lady Ranelagh as an executor of the will.[28]

From the date and contents of Gregory's memorandum it appears likely that Newton is referring to the projected Boyle lectures. His description of the "publick speech" and the contents of "ane Act," that is, a college speech, thesis, or disputation, closely resembles the contents of lectures given in 1692 by Bentley. Newton appears to be suggesting that his discoveries in celestial physics would serve the argument from design better than that reliance on the "contrivances" in animals and plants used by John Ray in his *The Wisdom of God Manifested in the Works of the Creation,* first published in 1691.

Newton's involvement with the Boyle lectures may have deepened when he had a chance to discuss his ideas with the trustees of the Boyle lectureship, with whom he met in early January at Boyle's funeral. Although Newton's attendance at Boyle's funeral has gone unnoticed, it is indicated in a letter of Samuel Pepys to John Evelyn, dated January 9, 1691/2:

[27] Turnbull, ed., III, 191. Henry Guerlac first brought this memorandum to my attention. Cf. Guerlac and Jacob, 307–318.
[28] Maddison, 15–16.

Sir,–I would have come to you the other night at St. Martin's on that grievous occasion [Boyle's funeral], but could not. Nor would I have failed in attending you before, to have condoled the losse of that great man, had I for some time beene in a condition of going abroad. Pray lett Dr Gale, Mr Newton and my selfe have the favour of your company to day, forasmuch as (Mr. Boyle being gone) wee shall want your helpe in thinking of a man in England fitt to bee sett up after him for our Peireskius, besides Mr. Evelin.[29]

Newton was in London at the time of Boyle's funeral, dutifully attended the services for his most eminent predecessor, and then met with Evelyn, Pepys, and Thomas Gale to discuss the events of the past week and to speculate about who might become Boyle's intellectual successor.

Three of the men at this gathering certainly knew of Boyle's intention to found a lectureship. Evelyn was one of its trustees; Newton, we are told by Gregory, had his own ideas for public speeches that were probably meant for the Boyle lectureship, and Pepys possessed a copy of the July 28 codicil that established the lectures.[30] The historian unaided by any known record of this meeting is left to muse about what may have taken place. Could it have been at this meeting that the name of the young Richard Bentley came up as the candidate most suited to give the first Boyle lectures? When we recall that Bentley had received from Newton in July 1691 directions for understanding the *Principia*, it is at least possible that Newton suggested Bentley as the clergyman best prepared to level an attack on atheists and materialists along the lines Newton had indicated to Gregory. We can only conclude on the basis of the available evidence that, whereas Evelyn and Tenison were most directly involved in the setting up of the Boyle lectures and in choosing Richard Bentley, Newton appears to have had a significant interest in this process.

[29] J. R. Tanner, ed., *Private Correspondence and Miscellaneous Papers of Samuel Pepys* (London, 1926), I, 51–52.
[30] Ibid., 48–49.

On February 13, 1691/2, Bentley was officially chosen as the first Boyle lecturer. His first sermon, delivered at St. Martin's in March, dealt with the role of religion in society. His second, at St. Mary-le-Bow, leveled an attack on materialism. In the third through fifth sermons, Bentley explicated the argument from design as it applies to the human body. On October 3, November 7, and December 5, 1692, the last sermons of that year, Bentley developed his version of the Newtonian natural philosophy. Before publishing these sermons Bentley consulted with Newton, and the first of Newton's four replies began with the now famous words: "When I wrote my treatise upon our Systeme I had an eye upon such Principles as might work with considering men for the beliefe of a Deity & nothing can rejoyce me more than to find it usefull for that purpose."[31] By way of assistance to Bentley, Newton may have written an account of his system of the world; part of what appears to have been a draft has turned up in his unpublished manuscripts.[32]

After delivering and publishing his sermons, Bentley rose in favor in the circle of the moderate churchmen. In his draft letter to William Wotton dated September 20, 1694, Evelyn commented about Bentley, "Be pleased then (when ever ye write to him) to let him know how ready I will always be to do him any service in my power."[33] A few months after this letter was written, Evelyn told Bentley exactly what he was willing to do for him. In November, Tenison and Evelyn had one of their usual meetings to discuss the Boyle lectures. At that time news came to them of the death of John Tillotson, archbishop of Canterbury. They then discussed the possible

[31] Turnbull, ed., III, 233. These letters were first published by Richard Cumberland, Bentley's grandson, in *Four Letters from Sir Isaac Newton to Doctor Bentley* (London, 1756).

[32] I. B. Cohen, "Isaac Newton's *Principia* . . . and Divine Providence," in S. Morgenbesser et al., eds., *Philosophy, Science and Method, Essays in Honor of Ernest Nagel* (New York, 1969), 542–547.

[33] MSS ADD 28104, ff. 18–19, B.L.

candidates for the position and agreed that Edward Stillingfleet should become the next archbishop.[34] He was already in
poor health, and so quite naturally the question of his successor arose. Evelyn wrote to Bentley, "We . . . immediately
voted the B. of Worcester [Stillingfleet] to succeed him
[Tillotson]; and after him (one day) Dr. Bentley: I am sure,
I augur it, expect it, wish it with all my heart."[35] Tenison and
Evelyn believed that the young Richard Bentley, classicist
and natural theologian, would crown his career as the primate
of all England. In their choice of Stillingfleet and then of
Bentley, Tenison and Evelyn were proved wrong. Tenison's
error may have been intended, since it was he who succeeded
Tillotson. In the years after 1694, Bentley made his career at
Cambridge. Yet as a Boyle lecturer he was in the position, had
he desired it, to seek considerable preferment within the
church.

Bentley's prospects are further evidence of the great importance attached by moderate churchmen to the Boyle lectures. In the early 1690s the lectures provided a podium from
which this moderate segment of the church could express its
concern over the post-Revolution order and in turn formulate
a philosophical structure to support its own concept of a
stable, religiously oriented society. The Boyle lecturers, in
particular the Newtonians, served as the spokesmen of the
church in this period.

This alliance of Newtonians and churchmen continued
throughout the reign of Anne. As party antagonism grew, the
ensuing split between low church and high church became
more extreme. By 1710 even social life within the church reflected this fundamental cleavage.[36] The diary of William
Wake[37] provides evidence of this division and also reveals the

[34] Evelyn MS, 2, v. I, f. 8, Christ Church.
[35] Ibid.
[36] G. Holmes, *British Politics in the Age of Anne* (New York,
1967), 21.
[37] See G. V. Bennett, "An Unpublished Diary of Archbishop William Wake," *Studies in Church History*, 3 (1966), 258–266.

extent of participation by the Newtonians in the low-church faction. They, and most of the Boyle lecturers, were in constant communication with Wake, who was an important assistant to Tenison and in 1716 succeeded him as archbishop of Canterbury. Among the frequent visitors to Wake's residence near St. James, Westminister, were Benjamin Hoadly, Samuel Clarke, Richard Bentley, John Harris, William Wotton, and Sir Isaac Newton.[38] Meeting at Wake's home was a group of low-churchmen and Newtonians that included either Newton or, more often, Samuel Clarke, with whom Newton was in close contact. In 1706, Newton met at Wake's home with White Kennett, who became a trustee of the Boyle lectures in 1711,[39] and Dr. John Hancock, who was Boyle lecturer for that year. The record provided by Wake reveals the existence in the reign of Anne of a coterie of moderate churchmen who centered around him and who included all the prominent Newtonians, most of the Boyle lecturers, and quite probably Newton himself.[40]

Apparently whenever they were in London they managed to visit Wake, and he was a constant source of information and communication among the low-church faction. By late in the reign of Anne, Newton's views on natural philosophy were so highly valued that churchmen outside his circle wrote to insiders, such as John Woodward, asking them "to hint anything

[38] For Hoadly, "Wake's Diary," April 18, 1706, Dec. 2, July 7, 1707, et seq.; Clarke's visits are too many to record here; Bentley, July 30, 1706, Aug. 13, 1707, Aug. 30, et seq.; Harris, May 4, 1706, Aug. 13, 1707, et seq.; Wotton, see Sept. 1706 and Aug. 1707, MS 1770, Lambeth Palace Library. Wake met twice with Newton in early November 1760. Also present on November 13 were Madame Gouvernet and Mr. Eachard.

[39] Currey & Co. MSS include a copy of the document nominating the new trustees. Aside from those above mentioned, they were Richard, earl of Burlington, Charles Trimnell, bishop of Norwich, Edmund Gibson, and Samuel Bradford.

[40] Whiston was not on good terms with Wake; see "Wake's Diary," Nov. 16, 1707. Among the other frequent visitors with Wake were Josiah Woodward and White Kennett.

briefly . . . in confutation or dislike of any late Undertaking or Hypothesis of Authors in Philosophy, and particularly of Sir I. N. etc. I shall have occasion to make use of it, when I'm glancing at ye Uncertainty and Precariousness of some Attempts in Philosophy."[41] In this period of low-church ascendancy the Newtonian natural philosophy as presented in the Boyle lectures was primarily an expression of latitudinarian thinking, and the circle that included the Finches, Wotton, Tenison, Evelyn, Bentley, Harris, later Samuel Clarke, and at all times Newton himself was the intellectual center of the church.

From this center emanated not only the main currents in religious thought, but also the channels of preferment. In a period when the church was deeply engaged in intellectual controversy, the Boyle lectureship naturally played a role in the system of church advancement. During the reign of William, and to a lesser extent of Anne, younger clerics came to rely for patronage on the good will of the moderate faction, and delivering the Boyle lectures was one sure road to advancement.[42]

In his directive establishing the lectureship, Boyle stated that its purpose should be to prove the Christian religion against "Atheists, Theists, Pagans, Jews, and Mahometans." Evelyn interpreted this statement as having been directed against "Atheists, [Deists], Libertines and Jews."[43] He apparently saw little need to attack theists and pagans, and in subsequent renderings of Boyle's intention the former term was dropped. The purpose of the lectureship was thus made

[41] MSS ADD 7647, f. 121, dated 1711, U.L.C.
[42] For interesting examples of patronage see MSS 953, f. 64–66, Lambeth Palace Library. Also on Stephen Nye's career, see Gibson MSS, 930, f. 56, Lambeth Palace Library. For an example of social mobility within the church based on intellectual merit see MSS ADD 4274, f. 126, May 21, B.L. The candidate is a Mr. Wilson. The career of Richard Bentley is also, to a lesser extent, an example of this phenomenon.
[43] deBeer, ed., V, 88.

more specific by its trustees and perhaps most clearly stated
by the first lecturer, Richard Bentley. He argued that in proving the Christian religion, the sermons "must be *contra malos
mores*, not *malos libros*."[44] Bentley was convinced that the
atheism of behavior and not the content of books constituted
the greatest threat to society. He too saw little need to attack
theists or Jews. Rather, his aim was to denounce the behavior
and philosophy that he associated primarily with Hobbism.[45]
To Bentley the purpose of the Boyle lectureship was to denounce a manner of political and social behavior that was
essentially Hobbist. The vision of man, the nature of social
relations, and the purpose of government presented in *Leviathan* bore a strange resemblance to the society within which
Bentley believed he lived. Indeed, the very settlement that
initiated the new order, contrary to the arguments about divine providence, rested on a fiat derived from the contract
between king and Parliament.

Bentley's quarrel was with the post-1688 order, and the
Boyle lectures were intended to reform it, to refute a growing
secularization represented by the "crafty, ill-principled men,"
and thereby to reform society in preparation for the final
reformation, possibly even for the coming of the new heaven
and the new earth. In the hands of the Newtonians, latitudinarian natural religion became refined into a social philosophy
intended to promote stability in the "world politick," and its
validity came to rest on the principles of Newton's natural
philosophy. The Boyle lectures created the Newtonian ideology as a justification for the pursuit of sober self-interest,
for a Christianized capitalist ethic. At first the lectures attacked mores and not books, but after 1695 and the lapsing of
the Licensing Act, the tension between Newtonians and freethinkers took on a sharper focus, as the deists and libertines
began, with the same force as in the 1650s, to answer back. They
launched their assault against the church, however, at the time

[44] Wordsworth, ed., I, 39.
[45] Ibid.

when its intellectual leadership possessed the skill and social cohesiveness to refashion the latitudinarian philosophy they inherited from the Restoration into a coherent and timely social and political ideology and to anchor it onto the model provided by Newton's system of the world.

CHAPTER 5

The Boyle Lectures and the
Social Meaning of Newtonianism

The lecture series endowed by Robert Boyle and administered in its early years by his close friends Evelyn and Tenison set the content and tone of English natural religion during the eighteenth century. By 1711 the reading of the Boyle lectures formed a part of an educated man's knowledge,[1] and of all the lectures those by Bentley (1692) and to a larger extent by Clarke (1704–05) and Derham (1711–12) exercised the greatest influence throughout Europe.[2] When d'Holbach attacked theism in 1770 he focused on the natural philosophy of the Boyle lectures,[3] and when Rousseau expressed his sense of God's presence in nature he enlisted Clarke as his primary defense.[4] Samuel Johnson learned much of what he knew about science from his reading of the Boyle lecturers.[5]

[1] See "Miscellanies of Anthony Hammond," Rawl. MS D. 1207, f. 23, Bodleian.

[2] Bentley's lectures were translated into Latin, German, French, and Dutch. See A. T. Bartholomew and J. W. Clarke, *Richard Bentley, D.D. A Bibliography of His Works and of All the Literature Called Forth by His Acts or His Writings* (Cambridge, 1908), 1–9. Derham's lectures went through thirteen English editions by 1768 and translations into Dutch, English, Swedish, and German. See John J. Dahm, "Science and Apologetics in the Early Boyle Lectures," *Church History*, 39 (1970), 4, n. 17. Clarke's lectures were translated into French in 1717.

[3] Paul Henri Thiry d'Holbach, *The System of Nature; or The Laws of the Moral and Physical World*, I (London, 1797), 35–78.

[4] *Émile*, trans., B. Foxley (New York, 1966), 231.

[5] Richard B. Schwartz, *Samuel Johnson and the New Science* (Madison, Wis., 1971), 106–107, passim.

The considerable reputation of these lectures in the eighteenth century is an indication of the moderate faction's
service to the church. The lecturers were carefully chosen by
the trustees, and they marshaled their arguments in defense of
natural and revealed religion with the conviction that their
efforts were critically important to the maintenance of the
church's moral leadership and political influence in a society
threatened at every turn by atheism. Both in retrospect and at
the time, the most significant intellectual achievement of the
Boyle lecturers in the period 1692–1714 was the integration of
Newtonian natural philosophy as the new underpinning of
liberal Protestant social ideology. This integration was mainly
the work of Bentley and Clarke and to a lesser extent of John
Harris, William Derham, and William Whiston.

Their lectures were read far more frequently by educated
men and women in the eighteenth century than were the scientific treatises of Newton. Yet despite their importance at the
time, the lectures of the Newtonians occupy only six years
out of the period of annual lectures that concerns us. Before
discussing the specifically Newtonian lectures, therefore, we
should know something about the content of the other lectures. They too tell us much about the intellectual concerns
and preoccupations of moderate churchmen after 1689, and
many of the themes explicated in the lectureship demonstrate
the continuity between those concerns and the ideas of Restoration latitudinarians discussed earlier.

In the common parlance of the seventeenth century, God
revealed his will through both his word and his work. The
lectures by the Newtonian commentators dealt with God's
work, specifically, with the operations of nature according to
the principles proclaimed by Sir Isaac Newton. An equally
strong theme in the Boyle lectures of the period centered,
however, on finding rational arguments to validate the truth
of God's word as revealed in Scripture. The lectures of John
Williams (1695), Samuel Bradford (1699), William Whiston
(1707), and Benjamin Ibbot (1713–14) stressed what they

called the rational and expedient credibility of the texts against what they claimed was a growing skepticism not only about the literal truth of certain prophetic passages, but also about the claim that the Scriptures were the work of God directly revealed to their writers. The motivation for these lectures appears to have been in part the fear that natural religion, if misunderstood or distorted, would destroy the supernatural content of Christianity.

John Williams, bishop of Chichester from 1696 and an ardent foe of such radical sectaries as the Muggletonians, claimed that debates about whether or not the sun is the center of the universe were hardly important compared to matters that affect our salvation. Belief in the divine inspiration of the Scriptures was one of those matters.[6] For Williams the question of the divine origin of Scripture was related intimately to God's ability to supersede, when necessary, the laws of nature. God seldom acts outside of nature,[7] but one such occasion had been his direct inspiration of the authors of Scripture. In maintaining the veracity of Scripture, Williams believed that he was asserting the providence of God.[8]

As we have seen, Anglican social and political theory hinged on this protean notion of God's providence. The latitudinarian churchmen could espouse their belief in social relations governed by the dictates of enlightened self-interest, maintain their defense of the Revolution of 1688–1689, and embrace the discoveries of the new science only if they could simultaneously hold together their model of the worlds political and natural by rendering all accountable to the dictates of providence. Williams apparently eschewed Thomas Bur-

---

[6] Williams, *The Possibility, Expediency and Necessity of Divine Revelation* (London, 1696), in S. Letsome and J. Nicholl, eds., *A Defense of Natural and Revealed Religion. . . . Sermons Preached at the Lecture Founded by the Honourable Robert Boyle*, I (London, 1739), 177–179; hereafter cited as Letsome and Nicholl.

[7] Ibid., 166–167.

[8] Ibid., 200.

net's embellishment of the Scriptural account of creation[9] in the *Sacred Theory* and deplored Spinoza's outright attack on the divine nature of the Scriptures[10] in his *Tractatus Theologico-Politicus* (1670) because each in his own way appeared to cast doubt on the literal account of divine efficacy in the universe. Williams and his fellow lecturers held to a very strict notion of exactly what sort of science and natural philosophy served their interests and those of the church. Only a mechanical universe operated at every turn by the divine will, making constant use, of course, of secondary causes, could maintain social and political order and stability. Similarly, every verse of Scripture reveals God's will; obedience to it, Williams assures us, will eventually result in the earth becoming "a kind of paradise again."[11]

Williams concludes his lectures with the promise of a millenarian paradise:

I freely acknowledge that there will be a great alteration in the present state of the Church, before the close of the whole, and before an end shall be put to Christ's mediatory kingdom upon earth: When "the Lord's House shall be established in the top of the mountain; and all nations shall flow unto it . . ." that there shall be but one Church over all the world, by the conversion of the Jews, and the coming in of the fulness of the Gentiles; and that by the coming down of the new Jerusalem from Heaven, it shall be in a state of perfect peace; and there shall be in that sense "a new Heaven and a new Earth." But that is a state in reserve; and there will need no evidence for that which is self-evident.[12]

Williams' defense of the veracity and credibility of Scripture entailed by necessity and by conviction an assertion of the church's millenarian dream of a future order dominated by Williams' distant ecclesiastical successors.

[9] Ibid., 191.
[10] Ibid., 180.
[11] Ibid., 242.
[12] Ibid., 246.

Williams remained vague on the means or methods by which this paradise would come into being. But in the lecture series of the previous year (1694), Richard Kidder, in the course of demonstrating the veracity of Christ's claim to be the Messiah, proclaimed that "the Apocalypse contains many Predictions, a great number of which are already fulfilled, and the rest are approaching; and we doubt not, but they will, in the due time, that is there prefixed and set down, be fulfilled also."[13] Kidder offers no predictions about time or place, but the warning he offers is unequivocal: God has promised horrible destruction if his subjects are sinful, and their demise, despite the concomitant promise of a millenarian paradise to follow in its wake, will be an event of tragic proportions.[14] Terror and devastation are not inevitable, but rather contingent on man's moral behavior. In contrast to Williams' rather benign assessment of the ease with which paradise may be instituted, Kidder frets over human sinfulness. He moralizes about the "carnal and worldly temper" of the Jews who rejected Christ and now resist conversion. Kidder's anti-Semitism obfuscates his message. In keeping with the casuistry of the day he means the Jews to represent the English nation, and clearly he, like so many other churchmen, is profoundly distressed by the moral climate of his age.

The Boyle lecturers were, however, never short of prescriptions for curing the atheism, immorality, and irreligion that apparently threatened to engulf them. Samuel Bradford convinced himself and possibly his listeners that many who reject Christianity are "such whose time is wasted in sport and luxury, who have never improved or exercised their higher faculties . . . nor furnished their heads with any solid materials to think upon."[15] He would cure laziness and irreligion

---

[13] Kidder, *A Demonstration of the Messias*, 3 vols. (London, 1694, 1699, 1700), in Letsome and Nicholl, I, 95.

[14] Ibid., 96.

[15] Bradford, *The Credibility of the Christian Revelation, from Its Intrinsick Evidence* (London, 1699), in Letsome and Nicholl, I, 437.

in a single effort: "Princes and great men of the earth may do very much to this purpose. . . . Others also, especially those bodies and societies of men, which have commerce with the Gentile world, might contrive methods for propagating their religion with their trade."[16] Bradford endorsed Boyle's desire, also proclaimed in his will, to spread Christianity to foreign parts and thereby to revitalize a militant Christianity, along with industry and trade, both at home and abroad.

Lecturing in 1707, Whiston prescribed a similar militancy. The spreading of the gospel to all nations would quicken the conversion of the Jews;[17] coupled with a renewed offensive against "that grand Mother of Idolatry, the great City of Rome, with its Tyrannical Empire"[18] this activism would bring closer the fulfillment of the prophecies. The openness of Whiston's millenarianism, displayed in his Boyle lectures and elsewhere, was never characteristic of the themes explored by the other lecturers. Most of their sermons, whether on the veracity of Scripture or on the necessity of natural and re- vealed religion, centered on the concerns of their present age.

In sermons intended for the Boyle lectureship but not de- livered because of ill health, William Fleetwood (1656–1723) linked the order found in nature with the expression of self- interest: "Natural causes still produce their natural effects, according to settled and establish'd laws of the creation; and men persue their inclinations and desires, according to their powers and opportunities, and every thing proceeds uninter- rupted in its usual, regular, expected course."[19] He warned, however, that all hinges on providence and not on secondary causes, which can, if necessary, be suspended.[20] Self-interest will work, provided God allows it to do so. Similarly Francis

[16] Ibid., 524.
[17] Whiston, *The Accomplishment of the Scripture Prophecies* (London, 1708), 182–183.
[18] Ibid., 218.
[19] Fleetwood, *An Essay upon Miracles* (London, 1701), 231.
[20] Ibid., 232–233.

Gastrell, lecturing in 1697, could only imagine the individual's freedom from arbitrary government and the "common liberty and equality of mankind" if that freedom rested on the authority of providence.[21] Yet paradoxically religion in the service of providence demands defense against its present-day attackers, because if "there is no God nor Religion . . . [then] all men are equal."[22] Without religion, it was believed, the hierarchical structure of society would crumble and with it property rights would disintegrate.[23]

With property and liberty at stake it is easy to understand why the Boyle lecturers bent every effort, marshaled every argument, and searched every Scriptural text for evidence with which to defeat the vanguard of irreligion: the deists, freethinkers, and atheists. Every lecture, on whatever topic, mentioned them specifically or generically. The final impression left to any reader of the lectures is that the latitudinarian segment of the church that controlled the Boyle lectureship and after 1689 was also the most influential faction within the church believed themselves to be living in a veritable state of seige, bedeviled at every turn by mores, values, thought, and behavior clearly in opposition to the church's interests, that is, to religion and piety. This conflict between the church and its opponents would appear arid and incomprehensible if we imagined it to be centered essentially on intellectual issues—can matter think? is motion inherent in matter? do chance and fate govern the operation of nature? These questions, hotly debated in sermon after sermon, symbolized the intellectual disagreement of diametrically opposed social attitudes and political positions. At this point the intellectual arguments marshaled by the church against the freethinkers concern us. Further along in this book the freethinkers will have their say.

Hobbes, Spinoza, Vanini, Bruno, Charles Blount, Toland,

[21] Gastrell, *The Certainty and Necessity of Religion in General* (London, 1697), in Letsome and Nicholl, I, 307.
[22] Ibid., 322.
[23] Ibid., 323.

Anthony Collins, and Matthew Tindal, even Epicurus, all
occupied a high place on our churchmen's list of offenders. In
their assault on Hobbes the Boyle lecturers relied heavily on
arguments developed during the Restoration. Indeed there was
hardly anything new that could be said against him, and those
arguments discussed earlier, in Chapter 1, do not warrant
repetition. At issue between Hobbes and the Boyle lecturers,
as with their latitudinarian predecessors, was the latter's perception of the Hobbist universe: mechanical laws governed by
material forces which complemented a social world of com-
peting interests dominated by the dictates of power and self-
interest. Place and fortune depended upon skill and not providence, and institutions survived according to how well they
served the marketplace.

The Boyle lecturers believed that Hobbes's prescriptions, as
they understood them, grew in popularity because he spoke to
contemporary circumstances and needs. Their task also was to
meet the needs of their age, to come to terms with a changed
social and political order, and to offer it a different version of
Christianity which, if accepted, would save men from destruction and preserve the church's interests.

By the late 1690s, however, Hobbes had acquired com-
panions in atheism. Spinoza was seen as one of the most
dangerous. His writings had circulated in England and, as
Rosalie Colie has documented, worked their influence in
deistic and freethinking circles, if not elsewhere.[24] The first
response to this challenge to Christian orthodoxy posed by
Spinoza came during the Restoration, from the Cambridge
Platonists, More and then Cudworth.[25] They did not share
Boyle's fascination for Spinoza, quietly expressed in letters
sent to the philosopher through Henry Oldenburg. Still
operating in the intellectual context of the 1650s, More at-

[24] Rosalie Colie, "Spinoza and the Early English Deists," *Journal of
the History of Ideas*, 20 (1959), 23–46.
[25] Rosalie Colie, "Spinoza in England, 1665–1730," *Proceedings of
the American Philosophical Society*, 107 (1963), 184–186.

tacked Spinoza's hylozoism as simply enthusiasm, akin to the beliefs of Quakers and Behmenists. Perhaps More knew about Spinoza's association with republican circles in Holland. Cudworth, on the other hand, imagined that Spinoza need not be taken seriously as a thinker.[26] Neither More's nor Cudworth's approach provided a sufficient rebuttal of Spinoza, and the Boyle lecturers found themselves faced with the task of yet again refuting his pantheism, which they generally misunderstood, and of denying the conclusions reached by his biblical criticism. Colie has elaborated amply upon the arguments brought to bear against Spinoza by John Williams, Francis Gastrell, John Harris, and most influentially by Samuel Clarke.[27]

Two of her points deserve emphasis. Increasingly the attack leveled against Spinoza focused on the practical, ethical implications of his pantheism. The freedom he awarded to men would allow them, in Gastrell's words, to live by "their own Determinations,"[28] in other words to pursue their interests, inclinations, or desires without regard to rewards or punishments externally imposed. Once again, as with Hobbes, Spinoza's system—although so profoundly different in character from the materialism and repression inherent in Hobbes's—in the eyes of churchmen amounted to nothing more or less than the same thing, a license for self-interest. And second, as the issue between Spinoza and the Boyle lecturers became not the veracity of the Scriptures but the philosophical and metaphysical foundations for human action, churchmen abandoned the tack of quoting the Bible at Spinoza and his followers and turned to natural philosophy and science for their arguments.

Clarke used Newton, and to a lesser extent Boyle, to argue for the infinity and omnipresence of the deity whose attributes do not include motion, but who regulates the motion of matter through the use of active forces or principles. Clarke

[26] Ibid., 210.
[27] Ibid., 204–210.
[28] Ibid., 205.

would keep the separation of creation from its creator while maintaining providence. The approach taken by the early eighteenth-century Boyle lecturers to the threat posed by Spinoza encapsulates an important but gradual transition that had occurred in Anglican thought since the Restoration. Because natural religion was intended to answer the moral problems posed by man's immediate material needs and not by his search for a spiritual and eternal destiny, it sought to validate itself inevitably on the physical order established by the natural philosophers. In the natural world governed by the laws of motion churchmen believed they had found a model toward which the political world could aspire. Arguments based on sacred history or Scripture—the validity of which the freethinkers were quick to deny—were hardly relevant to the worldly men Evelyn characterized as "crafty and ill-principled." The irony here is that in the process of combating Hobbes, Spinoza, and their supposed followers, latitudinarian churchmen eventually devised a version of Christianity dangerously similar to the ethical and philosophical deism they aimed to combat.[29]

Ironic too is all the publicity accorded by the Boyle lecturers to their deistic and freethinking opponents. Either by naming their adversaries directly and outlining their theories or by reference to them in marginal notes, the Boyle lecturers provided their listeners, readers, and subsequent historians with a fairly accurate account of the church's less vocal opposition. One of the most interesting sources of information for our purposes comes from the attack against the freethinkers leveled in the 1713 Boyle lectures of Benjamin Ibbot. He was responding to Anthony Collins' *Discourse of Freethinking* (1713), and his sermons are laced with references to a free-

[29] Through various approaches this same point has been made by other historians: ibid., 210; F. Wagner, "Church History and Secular History as Reflected by Newton and His Time," *History and Theory*, 8 (1969), 97–111.

thinking circle that espouses the doctrines of Giordano Bruno (1548–1600).[30] Recent inquiries into the activities and writings of Collins' circle reveal that Ibbot's charges were accurately aimed.[31] John Toland, Collins, and company, who did espouse the theories of Giordano Bruno, tended to look upon "the Gospel as a forgery or cunningly-devised Fable"[32] and regarded clergymen as the truly crafty and ill-principled. That Ibbot felt compelled to answer them charge by charge offers but one measure of the freethinkers' effectiveness.

Throughout all the intellectual cross-fire, controversy, and pamphleteering between churchmen and their freethinking opponents—a war waged most noisily and dramatically in the period from 1689 to 1730—the Boyle lectures of the Newtonians, more than those of other moderate churchmen, and the writings of Toland, Collins, and their friends strike at the heart of their respective positions and tell us most about their irreconcilable differences. To defend Anglican social hegemony and moral leadership the Boyle lectures of Bentley and Clarke and to a lesser extent of Harris, Whiston, and Derham refashioned latitudinarian social philosophy and rested it on the natural philosophy of Sir Isaac Newton, as they understood him. More effectively than any other body of literature the Boyle lectures provided the first and most durable explanation of Newtonianism available in the eighteenth century. As such they have been commented upon by historians of ideas as well as by philosophers. What I have to say about this early Newtonianism differs, in certain instances substantially, from what has been said elsewhere, and it may be useful to be explicit about these differences.

[30] Ibbot, *A Course of Sermons Preach'd for a Lecture Founded by the Hon. Robert Boyle, Esq. . . . Wherein the True Notion of the Exercise of Private Judgement, or Freethinking, in Matters of Religion is Stated* (London, 1727), 40–54, passim. Cf. M. C. Jacob, "John Toland and the Newtonian Ideology," *Journal of the Warburg and Courtauld Institutes*, 32 (1969), 325.

[31] See Chapter 6.

[32] Ibbot, 117.

Almost without exception the Newtonianism of this early period has been analyzed solely as an intellectual phenomenon, devoid of social and political content, transcendent of ideology. Within very recent historiography only Newton himself has fared slightly better as a social and political creature, subject to the concerns of his day and prepared to adjust his natural philosophy accordingly.[33] The "ism" of his followers (with which he substantially concurred[34]) and the sources of their opposition to the freethinkers emerge in recent writings, however, as solely religious or philosophical. These churchmen adopted Newton's natural philosophy, it is argued, because they feared the mechanical philosophy of Descartes and the support it would afford to atheists and materialists.[35] But why did they fear atheism and more particularly materialism if based on Cartesianism? Was it only simple Christian piety that was at stake? Was Newtonianism simply another weapon in the war between orthodoxy and heterodoxy?[36] That view renders religion as solely a matter of emotional conviction and

[33] See in particular Christopher Hill, "Newton and His Society," in Robert Palter, ed., The "Annus Mirabilis" of Sir Isaac Newton, 1666–1966 (Cambridge, Mass., 1970), 26–47; M. C. Jacob, "Toland," 321–324; and most briefly in P. M. Rattansi, "The Social Interpretation of Science in the Seventeenth Century," in Peter Mathias, ed., Science and Society, 1600–1900 (Cambridge, 1972), 1–32. There is the earlier, but often dismissed, B. Hessen, The Social and Economic Roots of Newton's Principia (Sydney, 1946). For all of its failings, at least it was a beginning.

[34] I do not wish to make light of the very real problems faced by historians engaged in discerning the subtleties and intricacies of Newton's thought. Neither will I accept the view that Newton stood aloof from the social and religious ideology that bears his name. For the latter view see William H. Austin, "Isaac Newton on Science and Religion," Journal of the History of Ideas, 31 (1970), 540–542.

[35] Herbert Drennon, "Newtonianism: Its Method, Theology and Metaphysics," Englische Studien, 68 (1933–1934), 397–409, is a useful and convenient summary.

[36] This seems a fair representation of the view taken by Dahm, 6; and G. R. Cragg, From Puritanism to the Age of Reason: A Study of Changes in Religious Thought within the Church of England, 1660–1700 (Cambridge, 1966, reprint), 107ff.

spiritual experience devoid of social reality. Or we are told that Newtonianism reasserted the nominalist and voluntarist strain of Christian theology developed in the late thirteenth century and that consequently "it is unnecessary, and, indeed misleading to postulate the influence of social and political analogies— for the influence was, if anything, exerted in the opposite direction."[37] It is extremely useful and important to recognize that the Newtonian natural philosophy belongs to a tradition of Christian thought, but it is a highly mechanical notion of intellectual change to assert that ideas are adopted simply because they are there to be adopted. Such an approach rests on the methodological assumption that the ideas of the Newtonians could and did exist independent of, or isolated from, a prevailing social, political, and economic environment.

Some historians have indeed recognized social environment but have misread the effect it had on the latitudinarian churchmen. To say that the Newtonians adopted the science of their master largely in response to "the presumed threat to religion of a science-supported atheism or deism"[38] ignores the ideological uses these churchmen then extracted from Newtonianism. It also presumes that the deists and atheists of this early period relied primarily on science for their arguments, rather than on philosophy or political theory, and that is highly presumptive. It would be truer to say that they sought support for their arguments wherever they could find it—in science, philosophy, political theory, popular culture, or in plain invective. At least one historian has seen that Anglicanism in the early eighteenth century expressed the interests and

[37] Francis Oakley, "Christian Theology and Newtonian Science: The Rise of the Concept of Laws of Nature," *Church History*, 30 (1961), 449.

[38] Robert E. Schofield, *Mechanism and Materialism: British Natural Philosophy in an Age of Reason* (Princeton, 1970), 21. Cf. P. M. Heimann, "Newtonian Natural Philosophy and the Scientific Revolution," *History of Science*, 11 (1973), 1–7.

desires of a ruling elite,[39] but having said that, we must know more precisely the problems faced by that elite and how it used Newtonian science as a weapon for explanation and self-justification. Yet to see natural religion supported by the new mechanical philosophy as an expression of social elitism is certainly more salutary than to imagine that because the proponents of the mechanical philosophy constituted a social minority they can be viewed as alienated intellectuals.[40] The anger and frustration of our churchmen at what they viewed as threats to the church's moral leadership and political power should never be confused with alienation in the modern sense of that word. They were, and they knew it, part of the natural leadership of the country, and they preached as they did because they wanted to control and harness change in such a way that it would serve and preserve their interests.

Because this essay is focused on the church's social and political interests I have inevitably paid less attention than is usual to the religious psychology and purely theological interests of the Newtonians and latitudinarians. In this regard Hélène Metzger's *Attraction universelle et religion naturelle chez quelques commentateurs anglais de Newton* (Paris, 1938) remains a classic study, although it must be supplemented by the vast literature of the last twenty years that deals largely with Newton's own science and natural philosophy.[41] This renaissance in Newtonian studies understandably has focused largely on the genius himself, almost to the exclusion of his followers. Yet the Boyle lectures by the Newtonian com-

[39] Roland Stromberg, *Religious Liberalism in Eighteenth Century England* (Oxford, 1954), chap. 10.

[40] Hugh Kearney, *Science and Change, 1500–1700* (London, 1971), 208.

[41] P. Casini, *L'universo-macchina: Origini della filosofia newtoniana* (Bari, 1969), represents essentially a distillation and summary of current research and is quite useful. Some of these comments were first made in my "Early Newtonianism," *History of Science,* 12 (1974), 142–146.

mentators provided their listeners and subsequent readers with
the first clear formulation of what became known as the Newtonian natural philosophy. Coupled with the more mathematical and technical commentaries of other Newtonians, Whiston, Henry Pemberton, Colin Maclaurin, and Jean-Theophile
Desaguliers, as well as with Newton's famous thirty-first
query to the *Opticks* (1717–1718), the Boyle lectures created
the Newtonian world view. Without their efforts Newton's
scientific accomplishments and his natural philosophy would
have remained unknown outside the small circle of his friends
or understandable only by those few scientists on either side
of the channel who could comprehend his mathematical and
scientific skills.

The churchmen who forged Newtonianism spoke not only
for themselves and the trustees of the Boyle lectureship, but
also for the latitudinarian faction within the church. Since the
early Restoration they had concerned themselves with finding
a comprehensive and liberal Protestantism that could address
the moral issues of their time. The threat to the church's political and religious power coming from the radicals had been
checked at the Restoration, and gradually their political influence all but disappeared.[42] Yet churchmen were not content. In the era prior to 1688–1689 they bemoaned the immorality of their age and sought to win their congregations to
the pursuit of enlightened self-interest as sanctioned by providence and confirmed by the laws of nature.

Latitudinarian natural religion was a well-developed creed
by the time of James II. Suddenly the church's interests were
once again threatened—now by monarchy itself. Some low-
churchmen turned to the prophecies for guidance and solace;
most sat on the sidelines as a revolution engineered by gentlemen and princes salvaged the church's hegemony.

As a result of the Revolution, an event of dubious morality
in the eyes of churchmen, the latitudinarians obtained control

[42] Christopher Hill, "Republicanism after the Restoration," *New
Left Review*, 3 (1960), 46–50.

over the church, but their position within the political establishment was ambiguous. The church had gained security, but
its political power and teaching authority rested on a precarious alliance with secular authority. At the Revolution government devised by contract and the necessities of self-interest
had triumphed and the latitudinarians were caught in a
dilemma. The Boyle lectures of the Newtonians accurately
reflect their plight. The natural religion of their Restoration
predecessors appeared to be the only solution. A liberal Christianity resting on the achievements of science, as obvious and
unassailable as they seemed, would secure allegiance to religion against the claims made by the Hobbists and freethinkers.
Loyalty to natural religion meant in the minds of low-churchmen loyalty to the church, and from that allegiance would
emerge social and economic behavior governed by providence.
Stability and prosperity would abound, and the millenarian
dreams of the church would be accomplished.

For some of the Newtonians, such as Whiston, this meant
the arrival of the literal "new heaven and new earth"[43] proclaimed by Scripture. For others such as Clarke, the dream
came to mean a stable and harmonious order governed by
providence where the necessities of the marketplace presented
no contradictions to the dictates of natural religion. The
Newtonians used the science of their master to support their
aims. Newton's natural philosophy served as an underpinning
for the social ideology developed by the church after the
Revolution.

Why did these churchmen so willingly embrace Newton's
scientific achievements and the natural philosophy that supported them? One possible answer is that Newton based his
system on scientific and mathematical calculations which were
essentially correct, and that Bentley and his colleagues under-

[43] Whiston, *The Accomplishment of Scripture Prophecies* (London,
1708), 95. Whiston sees the Newtonian natural philosophy as providing an explanation of natural events that renders the prophecies be-
lievable.

stood what Newton had done and thus were led inevitably to accept his natural philosophy. Yet it is doubtful that the first Newtonian commentators actually understood Newton's scientific and mathematical formulations. Bentley probably never managed to follow the intricacies of Newton's mathematics. Only Samuel Clarke, Boyle lecturer in 1704–1705, possessed the technical skill necessary to comprehend the *Principia*, and at the time of his lectures he was on intimate terms with Newton. Indeed Clarke's lectures rest on many of the same philosophical principles found in Newton's twenty-third query to the *Optice* (1706).[44] Even so they only treat in a more sophisticated and subtle manner the assumptions and arguments made by Bentley in 1692 when his acquaintance with Newton and his understanding of the technical aspects of the *Principia* were slight. Likewise the lectures of John Harris (1697) and William Derham (1713) follow the pattern set by Bentley and enriched by Clarke. There is no evidence to suggest that any of these lecturers received substantial tuition or direct indoctrination from Newton when they were actually involved in delivering their lectures. Although Newton appears to have had a hand in the Bentley lectureship he by no means masterminded the early public expositions of his natural philosophy.[45]

The conclusion becomes inescapable: if we are to discover why these churchmen adopted a world view based upon what they made of Newton, we must look to them for an answer. Their Boyle lectures reveal that they consciously rejected alternative natural philosophies in favor of the Newtonian system. What is more, they pitted this "new and invincible" system against the mechanical philosophy of Hobbes, the fatalism of Epicurus, the remnants of Aristotelianism, and the

[44] Repeated and enlarged into the thirty-first query to the *Opticks* (1717–1718). Clarke and his brother John were directly involved in the translation of the 1704 English edition into the 1706 Latin text.
[45] Cf. M. C. Jacob, "The Church and the Formulation of the Newtonian World-View," *Journal of European Studies*, 1 (1971), 130.

alternatives posed by contemporary freethinkers. The Newtonians did so not simply because they disagreed with these
explanations of how the universe worked, but because they
also and primarily saw these philosophies as profound threats
to the social, political, and religious order—the basis they
imagined for the security of church and state. The Newtonian
model provided for these churchmen a foundation upon which
that order might rest secure, and for that reason Newton's
philosophy proved irresistible.

The Newtonian churchmen, despite the original and highly
influential use to which they put Newton's system, were not
particularly original or creative thinkers. They had risen in
church circles through latitudinarian patronage, and many
of their arguments were culled from those of their, in some
cases more creative, predecessors, such as Barrow and Tillotson.
Whatever their limitations, however, by the late 1690s the
Newtonians were the intellectual leaders of the low-church
faction. Bentley rose through the patronage of Evelyn, Stillingfleet, and probably Newton. Samuel Clarke received comparable assistance from John Moore, bishop of Norwich. In
the late 1690s Clarke was introduced to him by William
Whiston, and Moore was very favorably impressed with
Clarke's understanding of Newton's discoveries.[46] Newton
knew Clarke from at least September 1697.[47] John Harris
probably received his appointment as Boyle lecturer through
the good offices of John Woodward, another Newtonian, to
Evelyn.[48] In the reign of Anne, Harris frequented the low-
church gatherings at the home of William Wake.[49] Derham's

[46] W. Whiston, *Historical Memoirs of the Life of Dr. Samuel
Clarke* (London, 1730), 5–8.
[47] MSS ADD 5, f. 239, U.L.C.
[48] Evelyn MS 39, f. 227. Christ Church. Woodward's manuscripts
are in the University Library, Cambridge.
[49] MS 1770, in 1706–1707, Lambeth Palace Library. Toward the end
of Anne's reign, Harris' sympathies may have turned away from the
moderates. He had gone to Wake for assistance and seems to have
been treated poorly; see entry for May 4, 1706.

affiliation is less clear. In 1689 he obtained a good place at an Essex parish; the timing of his move indicates that he probably had strong connections with influential moderates. By the early eighteenth century Derham was probably an acquaintance of Sir Isaac. At least in public and in contrast to their high-church colleagues, the Newtonians accepted the compromises of 1688–1689 and tended, by the reign of Anne, generally to support Whiggish political goals. Their social and political stances are in turn accurately portrayed in their natural religion.

In 1711–1712, William Derham lectured on the grandiose design ordained by God for all creation. In the opening lectures he demonstrates "what a noble Contrivance this [Gravity] is of keeping the several Globes of the Universe from shattering to Pieces, as they evidently must do in a little Time by their swift Rotation round their own Axes."[50] After his use of Newtonian arguments to demonstrate the providential design of the universe, Derham proceeds to argue for design in trees, insects, and even in English imperial exploits. He assures his readers that the spread of Christianity accomplished through English mercantile endeavors in China is a part of God's design. In contrast to the vicious attacks made by high-churchmen and Tory poets in this period on the unbridled greed of the moneyed men is Derham's assessment of the business world: "Thus the wise Governour of the World, hath taken Care for the Dispatch of Business. But then as too long Engagement about worldly Matters would take off Mens Minds from God and Divine Matters, so by this Reservation of every Seventh Day [Sunday], that great Inconvenience is prevented also."[51] In the above statement Derham brings the argument from design to its final, most useful conclusion. After the

[50] Derham, *Physico-Theology: or, A Demonstration of the Being and Attributes of God, from His Works of Creation* (London, 1714), 32–33.

[51] Ibid., 278, 437. Cf. I. Kramnick, *Bolingbroke and His Circle: The Politics of Nostalgia in the Age of Walpole* (Cambridge, Mass., 1968).

Hanoverian succession he was appointed to the canonry of Windsor.

Derham's assessment of the design inherent in the new social order is a shallow one by comparison to the elaborate and more interesting social ideology devised by the other Newtonian commentators. Since their thinking is similar, I shall treat their Boyle lectures as of a piece.

After 1688–1689 the formulation of the moderate church's social ideology rested on a new conception of the role of religion within the civil polity. Religion now exists in order to ensure the smooth running of the well-ordered society. The virtues instilled by religion, Harris claims, "do naturally and essentially conduce to the Well-being and Happiness of Mankind, to the mutual Support of Society and Commerce, and to the Ease, Peace, and Quiet of all Governments and Communities."[52] Religion supplies the appropriate means to the ends of self-interest, and the ends are questionable only when the means employed in their attainment violate the norms of Christian ethics.

Bentley expresses this new conception of the place of religion in the state when, in the course of his attack on Hobbes, he asks his opponent: "Why, then, dost thou endeavour to undermine this foundation, to undo this *cement of society*, and to reduce all once again to thy imaginary state of nature and original confusion? No community ever was or can be maintained, but upon the basis of religion."[53] As the social cement, religion binds in mutual attraction the various disparate forces that threaten the stability of society.

Resulting from the role now openly assigned to religion, the necessity arose to reorient church teachings toward the public sphere of human activity and away from the private matters of individual piety and worship of the creator. The

[52] John Harris, *Immorality and Pride, the Great Causes of Atheism*, in Letsome and Nicholl, I, 421.
[53] A. Dyce, ed., *The Works of Richard Bentley* (London, 1838), III, 22. Italics are my own.

practice of social virtue offers great reward to the individual, but more particularly, to society in general. Samuel Clarke admits to his listeners that this social reward is so obvious "that even the greatest enemies of all religion, who suppose it to be nothing more than a worldly or state-policy, do yet by that very supposition confess thus much concerning it."[54]

In order to reorient the aims of religion away from "otherworldiness," private devotion, or communication with the creator, and toward the formulation of a "state-policy," or the exercise of public and social virtue, religious thinkers had to place the highest possible value on the result obtained by such a reorientation. At the beginning of his lectures, Bentley presents the values promised by the social function of religion: "Religion itself gives us the greatest delights and advantages even in this life also, though there should prove in the event to be no resurrection to another. *Her ways are ways of pleasantness and all her paths are peace.*"[55]

Churchmen offered peace as the goal obtained by society when religion served as its support. To make credible their offer they enlisted Christ as the final authority for their attitudes. They claimed that he "hath enjoined us a *reasonable service*, accommodated to the rational part of our nature. All his laws are in themselves, abstracted from any consideration of recompense, conducing to the temporal interest of them that observe them. For what can be more availing to a man's health, or his credit, or estate, or security in this world, than charity and meekness, than honesty and diligence in his calling."[56] The sense of Bentley's statement is extremely important: Christ out of his own selflessness brought Christianity to

[54] Clarke, *A Discourse Concerning the Unalterable Obligation of Natural Religion, and the Truth and Certainty of the Christian Revelation,* in Richard Watson, ed., *A Collection of Theological Tracts* (1st ed., London, 1785), IV, 168.

[55] Dyce, ed., III, 13. Italics are Bentley's.

[56] Ibid., III, 19, conducing in 1st ed. "conducible." Cf. Watson, ed., 134.

men so they may ensure their temporal welfare by the pursuit of self-interest.

Religion condones and encourages a social and economic order based on the rights of private property and self-aggrandizement. It asks in return that men pursue their interests reasonably, that they observe the social virtues of justice and honesty. As Evelyn remarked in 1688, religious men do not abhor riches, they condemn only the *"vanity* of riches."* Clarke agreed that only the ruiners and spoilers "suffer themselves to be swayed by unaccountable arbitrary humours, and rash passions, by lusts, vanity and pride, by private interest, or present sensual pleasure, these, setting up their own unreasonable self-will in opposition to the nature and reason of things." The worldly men are "attempting to destroy that order, by which the universe subsists."[57]

Since churchmen were never very specific when they described the worldly minded, the historian can only be equally general in identifying those bugbears of the religious sensibility. The protest that began in the early 1690s against the moneyed classes became a flood of social criticism and satire in Augustan England. Much of this social criticism, especially by Tory wits such as Swift, Bolingbroke, and Gay, aimed at exposing the crude display of wealth and the vulgarity of manners typical of the new moneyed class. Its existence depended not on birth and often not even on the "laudable" acquisition accepted by Evelyn as a criterion for aristocratic stature. Rather, this new class assumed the social status of the lesser gentry and even the aristocracy largely because the great commercial and financial expansion of England after 1688 created a new social group based solely on its ability to invest wisely and to profit steadily. Their social mobility, coupled with a concomitant weakening of Restoration religious teachings and church influence, produced a new social order. Social

[57] Watson, ed., 133.

and political power no longer followed necessarily from one's high status in church circles, and religious sentiment that justified worldly success no longer ensured it.

The crafty, ill-principled men threatened the order of society and the place of religion within that order. Often they benefited at the expense of virtuous men, and as Clarke lamented, they created a situation which would "not only hinder [the virtuous] from enjoying those public benefits, which would naturally and regularly be the consequence of their virtue, but oft-times bring upon them the greatest temporal calamities, even for the sake of that very virtue."[58] Thus the delicate social order conceived by the Newtonians was continually jeopardized by the actions of irrational men bent on the ruthless pursuit of economic self-interest.

To ensure the order of society, religious men sought to create a model for its workings. They believed it necessary to expound this model publicly because the

corruptness of the present estate which human nature is in, the generality of men must not by any means be left wholly to the workings of their own minds, to the use of their natural faculties, and to the bare convictions of their own reason; but must be particularly taught and instructed in their duty, must have the motives of it frequently and strongly pressed and inculcated upon them with great weight and authority, and must have many extraordinary assistances afforded them; to keep them effectually in the practice of the great and plainest duties of religion.[59]

In an effort to correct the corruptness of men, to exhort them frequently and strongly "with great weight and authority," churchmen turned to the study of physical nature. Before discussing the important lessons they drew from nature and the application made of those lessons to society we must pause to consider the question of why so many churchmen in the late seventeenth and early eighteenth centuries turned to

[58] Ibid., 174.
[59] Ibid., 189.

scientific study, an endeavor that contributed so decisively to the respectable development of modern science and technology.

A partial answer is provided for us by Samuel Clarke, probably one of the few churchmen in England who genuinely understood Newtonian physics and who, in turn, based his entire philosophical and ethical thinking on the physical concepts underlying Newton's scientific achievements.[60] As has been argued, English religious thinkers in the late seventeenth century came to assign meaning to religion insofar as it served a social function. Religion performed the task of contributing order to society, of checking greed and avarice, or ensuring stability within the society it served to cement. Concomitant with the new emphasis placed on the social role of religion was the importance awarded to God's providence in the universe and in the affairs of men. God directs the operations of both realms; it is his "wisdom and justice and goodness in the disposition and government of the moral world, which necessarily depends on the connexion and issue of the whole scheme."[61] Confidently, Clarke asserts that God's providence operates in both the world natural and the "world politick."

Yet Clarke reveals his concern, and a vexing and troubling one it must have been, that God's providential action in the "world politick" is hidden from us, that we see only dimly the effects of his providence in our affairs. But in the natural order God is clear and distinct; he is as clear as the logic with which Clarke explicates his attributes and effects.

It may here at first sight [Clarke writes] seem to be a very strange thing that through the system of nature in the material, in the inanimate, in the irrational part of the creation, every single thing should have in itself so many and so obvious, so evident and undeniable marks of the infinitely accurate skill and wisdom of their almighty creator, that from the brightest star in the firmament of heaven, to the meanest pebble upon the face of the earth, there

[60] See John Gay, "The Idea of Freedom as the Basis of the Thought of Samuel Clarke" (Ph.D. diss., Columbia University, 1958).
[61] Watson, ed., 178.

is no one piece of matter which does not afford such instances of admirable artifice and exact proportion and contrivance, as exceeds all the wit of man (I do not say to imitate, but even) ever to be able fully to search out and comprehend; and yet, that in the management of the rational and moral world, for the sake of which all the rest was created, and is preserved only to be subservient to it, there should not in many ages be plain evidences enough, either of the wisdom, or of the justice and goodness of God, or of so much as the interposition of his divine Providence at all, to convince mankind clearly and generally of the world's being under his immediate care, inspection and government.[62]

The evidence of God's efficacy in the moral and political order would be present only after "the period and accomplishment of certain great revolutions."[63] That is, man's complete dependence upon God will become fully evident even to the most uninhibited atheist when the divine plan in history is fulfilled and the new heaven and the new earth are accomplished. Until such time, men must simply believe in God's providential action in the "world politick." But when men are searching to interpret the implications of the providential design for the nature of society, faith is simply not enough. If God's plan in the natural world can be understood, and nature does serve as a guide or model for the processes of the moral world, then in the study of physical nature may be found an alternative to the corruptness and wickedness of "this vicious age." By the late seventeenth century, science provided clearer evidence than did society for the efficacy of providence, and churchmen were drawn to the study of natural philosophy in increasingly large numbers.

As an alternative to "this vicious age," Clarke presents in the Boyle lectures the conclusions drawn from his study of nature. The physical principles and laws of motion that express the fiat of God and comprise the structure of the New-

[62] Ibid., 177–178.
[63] Ibid., 178.

tonian natural philosophy served for Clarke to explicate the nature of the social order.

On the surface it might appear that concepts such as the void, absolute space and time, matter and motion, bear little relevance to a model of society desired by churchmen. Yet the very philosophy of nature that lay beneath Newton's mathematical and experimental endeavors,[64] and was in turn adopted by his commentators, arose from an intellectual and social milieu that gave even notions about physical nature an ideological significance. The notion that matter is dead or lifeless received its most rigorous explication by the Cambridge Platonists, Henry More and Ralph Cudworth. They argued their case in the face of mid-century materialists and atheists who asserted a life-principle in matter and thereby proposed the mortality of the soul. These ostensibly philosophical notions were integral to the radical social and political aims of certain sects such as the Levellers. Newton, of course, drew much of his natural philosophy from the Cambridge Platonists. I do not wish at this time to discuss the complexities of these political and ideological controversies of mid-century. It is necessary only to point out that even in its mid-century origin the natural philosophy of Newton and his commentators was never divorced from its social context.

The notion that matter is dead or lifeless, extended and impenetrable, is fundamental to the Newtonian natural philosophy. Whether in the form of a planet or in its elemental atomic state, matter is moved only by an outside, immaterial force. The source of motion is God. He is the origin and constant source of the motion present in the universe. Newton expresses his concept of matter most clearly in the thirty-first query to the *Opticks* (1717–1718): "The *Vis inertiae* is a passive Principle by which Bodies persist in their Motion or Rest,

[64] See Henry Guerlac, *Newton et Épicure, Conférence donnée au Palais de la Decouverte le 2 Mars 1963*, Paris, 1963; and my opening remarks in Chapter 1.

receive Motion in proportion to the Force impressing it, and resist as much as they are resisted. By this Principle alone there never could have been any Motion in the World."[65]

The laws of motion, such as universal attraction, operate on matter at a distance. The void is essential to this operation. The space within which matter is attracted and moved is an entity unto itself; within that entity God's power in the universe operates. In effect, the law of universal attraction is the will of God expressed in the universe,[66] and its "admirable order" manifests the providential will of an intelligent and all-powerful being.

After the Revolution, its supporters argued that this same providential will had brought about the events of 1688–1689. In keeping with this providentialist theology the Newtonian commentators emphasized God's role in the universe. At creation, God put the atoms of the universe into order; out of the primeval chaos God constituted the "present frame of things."[67] This frame is maintained only because of God's constant intervention. "Consequently [Clarke tells us] there is no such thing as what men commonly call the course of nature, or the power of nature. The course of nature, truly and properly speaking, is nothing else but the will of God producing certain effects in a continued, regular, constant, and uniform manner; which course or manner of acting, being in every moment perfectly arbitrary, is as easy to be altered at any time as to be preserved."[68] God as a pure, spiritual being, immense, omnipresent, full, and all-powerful, exercises complete authority over matter. He is the final source of all motion in the universe; and he may delegate this power to man.[69]

[65] Newton, *Opticks* (2d ed. with additions, London, 1718), 372–373.

[66] Samuel Clarke, *A Demonstration of the Being and Attributes of God* (London, 1705), 69–70, 86–87; reprinted Stuttgart-Bad Cannstatt, 1964.

[67] Dyce, ed., 148–149.

[68] Watson, ed., 246.

[69] Ibid.

In the Newtonian system, the power given to man to regulate, to move, or to change matter and ultimately the course of nature[70] vindicates latitudinarian social teachings. Matter is "brute and stupid," and man has ultimate control, by virtue of his reason, over the things of the world. Bentley finds it absurd to imagine "that atoms can invent arts and sciences, can institute society and government, can make leagues, and confederacies, can devise methods of peace and stratagems of war"[71] or could "transact all public and private affairs, by sea and by land, in houses of parliament, and closets of princes."[72] The Newtonian definition of the relationship between man and matter gives a philosophical sanction to the pursuit of material ends, to using the things of this world to one's advantage, in effect, to bargain, to sell, to engage in worldly affairs with the knowledge that this activity is a God-given right. Man's power to acquire material possessions is an extension of his right to change the very course of nature, to control "brute and stupid" matter. But in his material dealings, man must instill an order similar to that imposed on nature by God. Just as God is continually promoting the universal benefit of the whole, men must also engage in universal benevolence. Man is "obliged to obey and submit to his superiors in all just and right things, for the preservation of society."[73]

In the process of controlling matter, men are assigned certain stations. It is essential for the preservation of social order that the individual "attend the duties of that particular station or condition of life, whatsoever it be, wherein providence has at present placed him; with diligence, and contentment; without being either uneasy and discontented, that others are

[70] Ibid.
[71] Dyce, ed., 49–50. In this same passage Bentley makes a derogatory reference to the discourses of hinds and panthers. He is referring to John Dryden. The Boyle lectures did not always avoid mention of controversies among Christians themselves. Note the similarity of these passages to statements by Barrow quoted in Chapter 1.
[73] Watson, ed., 122.

placed by providence in different and superior stations in the world; or so extremely and unreasonably solicitous to change his state for the future, as thereby to neglect his present duty."[74] The social structure is sanctioned by God; it is singularly perverse to attempt its alteration.

Such an attempt was made during the Interregnum when "men of ambitious and turbulent spirits, that were dissatisfied and uneasy with privacy and retirement, were allowed by his [Epicurus'] own principles to engage in matters of state."[75] Bentley refers to the divisive sectaries who, by their inroads into the power of the aristocracy, men of "privacy and retirement," and the court, undermined the political and social order. Bentley describes their actions as Epicurean and atheistical, attitudes that he discerns in post-1688 society.

Likewise, Samuel Clarke makes constant mention of the corruption and disorder infecting the world around him: "The condition of men in this present state is such, that the natural order of things in this world is in event manifestly perverted, and virtue and goodness are visibly prevented in great measure from obtaining their proper and due effects in establishing men's happiness proportionable to their behavior and practice."[76] The pessimism produced by this insight into the worldly conduct of men who ignore God's providence and government of affairs leads Clarke, as it led other contemporary churchmen, to associate the perversion of social values with the decay of nature. The analogy is ever-present in his sermons:

The sun's forsaking that equal course, which now by diffusing gentle warmth and light cherishes and invigorates everything in a due proportion through the whole system; and on the contrary, his burning up, by an irregular and disorderly motion, some of the orbs with insupportable heat, and leaving others to perish in extreme cold and darkness: what this, I say would be to the

[74] Ibid., 148.
[75] Dyce, ed., 24.
[76] Watson, ed., 111–112.

natural world; that very same thing, injustice and tyranny, in-
iquity and all wickedness, is to the moral and rational part of the
creation.[77]

Clarke believed that the decay of moral values and consequently of social stability prefigured the physical decay of
nature. If the destruction of nature is to be prevented, then the
decay of moral values must cease. The principles of order and
reason demand assertion, and to do this, the order inherent in
nature must be affirmed and applied to the moral and social
order. Thus churchmen turned to the Newtonian system as
the model inherent in nature and applied its principles to the
society within which they lived.

Gravity, the universal and mutual attraction of all bodies
in the universe, is the ordering principle of nature. It is the
fiat of God operating as the laws of gravitation which instills
order and harmony in the universe. Gravity acts on matter in
a void, and because of this vacuum it is possible for the atoms
to be moved into place, to be formed by gravity, into the
matter and form of the universe. The power of mutual attraction is "a new and invincible argument for the being of
God."[78] It is the evidence of mutual attraction in the universe
which gives final affirmation to God's providential activity in
nature. Clarke bases his assertion of God's providence on this
argument:

that most universal principle of gravitation itself, the spring of
almost all the great and regular inanimate motions in the world,
answering . . . not at all to the surfaces of bodies (by which
alone they can act one upon another), but entirely to their solid
content, cannot possibly be the result of any motion originally
impressed on matter, but must of necessity be caused (either immediately or mediately) by something which penetrates the very
solid substances of all bodies, and continually puts forth in them
a force or power entirely different from that by which matter

[77] Ibid., 143.
[78] Dyce, ed., 157–167, esp. 158, 163.

acts on matter. Which is, by the way, an evident demonstration, not only of the world's being made originally by a supreme intelligent cause; but moreover that it depends every moment on some superior being, for the preservation of its frame. . . . Which preserving and governing power, whether it be immediately the power and action of the same supreme cause that created the world, of him "without whom not a sparrow falls to the ground, and with whom the very hairs of our head are all numbered"; or whether it be the action of some subordinate instruments appointed by him to direct and preside respectively over certain parts thereof; does either way equally give us a very noble idea of providence.[79]

The physical principles explained mathematically by Newton in the *Principia* offered to churchmen what appeared to them as undeniable proof of God's providence. The principle of universal gravitation presented a means whereby the decay of nature might be avoided; it became the basis of a system "for keeping the several Globes of the universe from shattering to Pieces."[80] An alternative could be offered to Evelyn's fears that this world would go up "like a bomb" unless we accomplish the reformation. The Newtonian commentators were never simply intent on explicating a principle that would ensure the stability of the natural order. Their system only began with the physical principles explained in the *Principia*. They knew that ultimately events in the natural realm hinged directly upon events in the "world politick." Their aim was to construct both a physical and moral model, a social ideology grounded upon Newton's science, that would bring about the reformation.

Clarke makes manifestly clear the purpose of the system he constructs on the basis of the universal and mutual attraction of all bodies: "All inanimate and all irrational beings, by the necessity of their nature, constantly obey the laws of their creation; and tend regularly to the ends for which they were

[79] Watson, ed., 116–117.
[80] Derham, 32–33.

appointed. How monstrous then is it, that reasonable creatures, merely because they are not necessitated should abuse that glorious privilege of liberty, by which they are exalted in dignity above the rest of God's creation, to make themselves the alone unreasonable and disorderly part of the universe!"[81]

The disobedience of rational creatures, endowed by God with the power of moving matter and thereby defining their freedom through action, has worked against God's will and thereby blurred our vision of God's providential action in the moral order. But with the pattern of God's providence and benevolence in the world natural revealed by a study of Newtonian physics, Clarke returns to the "world politick" and argues for the insertion of a comparable order based on the obedience of rational creatures to the providential will of God. The reasonableness of the natural world must be effected in the civil polity: "And the practice of universal justice, equity, and benevolence, is manifestly . . . as direct and adequate a means to promote the general welfare and happiness of men in society, as any physical motion or geometrical operation is to produce its natural effect."[82]

With a puritanism common to moderate churchmen, the explicators of Newton catalogued the impiety and profanity of their age. They considered "what vast Loads of Filth, of all Kinds, are to be seen up and down in Heaps amongst us. Atheism and Deism, Scepticism and Infidelity, Immorality and Profaneness, often Contempt of God and all Religion; and prostituting of all things to private Profit and Advantage."[83] This bleak characterization by Harris aptly expresses sentiments common in church circles after the Revolution of 1688. The "possessive individualism"[84] seen by modern historians in

[81] Watson, ed., 140.

[82] Ibid., 173. Cf. Dyce, ed., 174–175.

[83] J. Harris, *The Practice of Religious and Moral Duties, the Best Way to Make a Nation Happy* (London, 1701), 13. Cf. Watson, ed., 112.

[84] C. B. Macpherson, *The Political Theory of Possessive Individualism, Hobbes to Locke* (Oxford, 1962).

the writings of Locke and Hobbes had become a dominant aspect of social behavior in Augustan England. Yet in their criticism of this behavior, moderate churchmen did not simply condemn the pursuit of self-interest. In contrast to their high-church peers, the Newtonians offered justification for a certain style to be used in the pursuit of one's self-interest. They condemned only those who separate private interest from public interest, those who "find fault with each other's management, and, through self-conceit, bring in continual innovations and distractions."[85] The disorderly pursuit of self-interest disturbed Newton's followers. In society men must assume a power and direction over matter similar to the power exercised by the divine fiat in nature. As the only other beings endowed with the power of moving matter, they must impose an order and pattern in the exercise of their prerogative.

Man must acquire in his conduct of worldly affairs the same overriding reason evident in God's operation of the physical universe. Just as God's absolute power is never arbitrary because of his eminent reason, the power of men over matter, their power to mold the inanimate to suit their needs, must be controlled by the cultivation of reason. The harmonious effects of God's reason are natural to the universe; the efficacy of reasonable men, "Truth Justice, and Benevolence, do naturally and essentially conduce to the Well-being and Happiness of Mankind, to the mutual Support of Society and Commerce, and to the Ease, Peace, and Quiet of all Governments and Communities."[86]

In the arguments used by Newtonian commentators, a new definition of social virtue emerges. The right of the individual to pursue his self-interest is affirmed; the preservation of one's interests is analogous to the process of self-preservation inherent in the natural order. This process always works for the benefit of the whole; nature is integral and harmonious. Like-

[85] Watson, ed., 162.
[86] Harris, in Letsome and Nicholl, I, 421.

wise, private interest must bend to public necessity. The obedience found in the inanimate order is a model upon which the dictates of self-interest must be tempered by the needs of society. The desire for power is natural, but its misuse is unnatural. Thus social virtue becomes "natural" to man, and antisocial behavior renders him aberrant or "unnatural."[87] The Newtonian social ideology primarily deemed man's social sinfulness unnatural.

Man's power over matter places him in a relationship with society analogous to God's relationship with the universe. Just as the divine goodness regulates the course of nature and promotes the universal benefit of the whole, man must also engage in universal social benevolence. His power to regulate the course of things enables him to structure society such that its workings produce a universal harmony.[88]

The desire for social harmony common to all the early commentators on Newton, and indeed to the entire moderate faction of the church, led them to subdue even religious emotion to the needs of the ordered society. Out of zeal for social harmony, Clarke argued that social virtue must arise from "a frequent and habitual contemplating [of] the infinitely excellent perfections of the all-mighty creator and the all-wise governor of the world."[89] The study of the physical order leads to social virtue, and "a due subjecting [of] all our appetites and passions to the government of sober and modest reason [is] the directest means to obtain such settled peace and solid satisfaction of mind as is the first foundation and the principal and most necessary ingredient of all true happiness."[90] Religious feeling finds its most socially useful expression through the systematic study or contemplation of nature. All other human emotion must similarly be controlled

[87] Watson, ed., 162, 140, 161.
[88] Ibid., 121–122.
[89] Ibid., 172.
[90] Ibid. See Maren-Sofie Røstvig, *The Happy Man. Studies in the Metamorphoses of a Classical Ideal*, II (Oslo, 1958), 166–167.

by reason. In the sober world of the Newtonian commentators, science and natural philosophy flourish along with piety and moderation, and the irrational in man is repressed to fit the needs of the harmonious society.

This system, wherein individual endeavor is socially directed and arises from an essentially religious motivation, and wherein human needs are met only when their satisfaction suits the needs of society, receives its final sanction from a providential God. He presides over the modern world. His subjects do his work because it suits their self-interest; they subdue their power to the needs of the whole. Their reward consists in social harmony and in the knowledge that their behavior is "natural." Its significance is cosmic; when acting reasonably man complements the natural order. The resulting complacency is in turn augmented by a study of nature, or science, for it reassures men that the natural model remains intact.

But the promised complacency of the Newtonian vision seldom entered the troubled minds of the churchmen who promoted it. Before complacency would be theirs they would first have to do battle with the Hobbists, atheists, and Epicureans who threatened the very fabric of the society churchmen envisioned.

The primary purpose of the Boyle lectures, as conceived by their benefactor, Robert Boyle, had been to attack deism, atheism, and other heresies that threatened the Christian religion. The churchmen who gave the lectures followed his command. Yet the Boyle lectures of the Newtonian commentators, and indeed of all the lecturers, attempted to do much more than simply refute the heretics. The lecturers offered a new social philosophy—one that was tied intimately to their understanding of the physical order. Only then did they turn their attention to the arguments presented by their opponents.

Many freethinking arguments were of ancient origin. Those of Lucretius, revived in England during the mid-seventeenth century, are but one example. The heretical protagonists of the latter part of the century often repeated the arguments of their mid-century elders, of whom even in the 1690s Hobbes remained the doyen. Repetition was never confined to the ungodly, and our churchmen repeated the arguments used by Boyle, More, Cudworth, Walter Charleton, and a host of less-known commentators. Indeed the lineage of these attacks and counterattacks is worthy of a separate study.

Bentley labeled atheism as folly, but a form of it that was a dangerous threat to society. He addressed Hobbes with the question quoted earlier. "Why, then, dost thou endeavour to undermine this foundation, to undo this cement of society, and to reduce all once again to thy imaginary state of nature and original confusion?"[91] Hobbist principles and men who act on them threaten the order and stability of society, and Bentley believed their views to be so widespread that he warned, "There's too much reason to fear, that some of all orders of men, even magistracy itself, have taken the infection, a thing of dreadful consequence, and most imminent danger."[92]

If atheism is tolerated in high places, then the result will be arbitrary government and tyrannical oppression. On the basis of Hobbes's principles, men would not be held responsible for their actions.[93] Social virtue would cease as a force operative in society. To deny God's existence, or to assert that God and matter are one, would in its turn deny man's power over matter, that is, his free will. The power of men to pursue their self-interest would cease, and larger social forces would take precedence over the interests and power of individual enterprise.

On the basis of Epicurean principles these same social forces would triumph. Their manipulators would be atomlike men, who blindly collide and tumble as they seek power and preferment. Bentley decries a society where "atoms can invent arts and sciences, can institute society and government, can make leagues, and confederacies, can devise methods of peace

[91] Dyce, ed., 22.
[92] Ibid., 23–24.
[93] Harris, in Letsome and Nicholl, I, 433.

and stratagems of war."[94] Yet it appears that the Epicureans flourish with a vigor second only to that of the Hobbists. Even cautious gentlemen "can wink and swallow down this sottish opinion about percipient atoms, which exceeds in incredibility all the fictions of Aesop's fables."[95]

Surrounded as they thought by ungodly men bent on capitalizing on the new order, churchmen turned to a direct confrontation with the followers of Hobbes, Epicurus, and the like. Perhaps the most interesting argument of the many devised is that offered by Samuel Clarke in his critique of Hobbes. Clarke accepts the notion of a contract or compact as fundamental to the workings of society. With a realism often uncommon to churchmen, he suggests that he has recognized the nature of the 1689 settlement. But Clarke refuses to accept the vision of man's nature that lay at the basis of Hobbes's contract theory. With Hobbes, men devise their own laws to fit the necessity of their situation. These "positive laws" receive their sanction from the pressures and forces operative within society, and not as Clarke would have it, from "the nature of things." In Clarke's terms, civil government receives its ultimate legitimacy from the "laws of God and nature" which are "the eternal reason and unalterable nature and relations of things themselves."[96] The providential God decrees an order or course of nature that conforms to certain discernible physical and mathematical laws. Similarly his providence and authority are the foundations upon which the civil polity rests.

In the order ordained by Newton's God, the Hobbist description of the state of nature becomes an absurdity. Clarke argues that if men had been corrupt and brutal in their original state, the contract necessary for the preservation of order and society would never have arisen. He posits therefore a state of

[94] Dyce, ed., 49–50.
[95] Ibid., 49. This comment also indicates Bentley's early interest in the ancients-and-moderns controversy.
[96] Watson, ed., 157–159.

nature directly opposite to that invoked by Hobbes. In Clarke's conception, men lived in an original state governed by the rules of natural religion, by justice, equity, and universal benevolence. Thus, reason is the "proper nature of man, [and] can never . . . lead men to anything else than universal love and benevolence: and wars, hatred, and violence can never arise but from extreme corruption."[97] Sin is unnatural; and behavior that is disruptive to society is not simply wrong, it too is unnatural.

Clarke is able to apply this rigid criterion to human behavior because he rests his case on a Newtonian understanding of the physical world and God's government of it. Clarke saw nature as working harmoniously because matter is controlled by a law of universal attraction. Operative through the will of God, for whom space exists as a sensorium, universal attraction maintains the course of the planets and causes the atoms to come together in the void. Their configurations form the matter of our world. By God's decree, man has power over this matter. This power, his free will, gives man a mandate to manipulate material things, to engage in commerce, to conduct affairs of state. The harmony imposed by God in the world natural confers direction on man's power in the "world politick." He, too, is bound to impose harmony. Men "should likewise employ those their extraordinary faculties in preserving the order and harmony of the creation, and not in introducing disorder and confusion therein."[98] To do otherwise would be to act unnaturally.

Yet churchmen saw moral disorder and political confusion rife in their society. And because they believed in the analogy between the "world politick" and the world natural, they often feared that should disorder triumph in their world, then destruction became imminent in the natural order. The Newtonians primarily sought to present a vision of society that would nurture stability and harmony and would also guaran-

[97] Ibid., 160–161.
[98] Ibid., 167.

tee a Christianized pursuit of the individual's power and duty in the social order. But to secure that vision the secular-minded who ruthlessly pursued power and preferment, and who would undermine the foundation of all government and authority, had to be stopped. By their denial of God's providence and their advocacy of Hobbist or Epicurean notions, the libertine and atheistical would constitute society on terms that excluded religion as a political or social force. They would undermine the position of the church. In so doing, they would set society on a course that could only lead to its destruction. The Newtonians never ceased to fear the designs of worldly men who, they imagined, would constitute society and government along purely secular lines.

What the moderate churchmen never realized was that their social teachings, based on the formidable order of the Newtonian universe, offered a powerful justification for the very order that disturbed them. For in the market society that flourished in eighteenth-century England under the sanction of God's providential design, even the "crafty, ill-principled" men eventually found a place, albeit unwittingly, in the new and grand design.

CHAPTER 6

The Opposition:
Freethinkers

The Revolution of 1688–1689 undid more than the Stuarts. While it instituted a constitutional system that ultimately produced political stability and the domination of government by what Disraeli first called the "Venetian oligarchy," the institution of a comparable intellectual conformity or repression eluded either church or state. The irascible Humphrey Prideaux complained that toleration "hath almost undone us, not in increasing the number of dissenters but of wicked and profane persons; for it is now difficult almost to get any to church, all pleading the license, although they make use of it only for the ale house."[1] The wicked and profane outside of court circles had been largely mute during the Restoration;[2] strict licensing laws had curtailed the publication of dangerous works. In May 1695 the Licensing Act lapsed and was not renewed. Even before that year intellectual controversy and opinion of a totally unorthodox nature had emerged—the antitrinitarian controversy of the early 1690s being but one example[3]—but after 1695 the dikes were broken by a series

[1] *Letters of Humphrey Prideaux . . . to John Ellis, Sometime Under-Secretary of State, 1674–1722* (London, 1875), 154, June 27, 1692.
[2] But mention of atheism did not cease; of particular interest, [A Person of Honour], *The Atheist Unmasked . . .* (London, 1685), or in a lighter vein T. Otway's play, *The Atheist: Or, The Second Part of the Souldiers Fortune* (London, 1684).
[3] In 1693, White Kennett noted that "atheism drives furiously on," Lansdowne 1013, f. 51 v, B.L. Socinian pamphlets were being privately circulated in 1689; MSS ADD Hatton-Finch 29573, ff. 355–370, B.L.-

of freethinking tracts, poems, and lampoons, most of which were aimed against the church.[4]

After 1689 the Newtonians continued the church's assault on Hobbes, but by the end of the century he had become more of a symbol than a reality. The Newtonians did use their master's natural philosophy against him; their attitude toward Hobbes is one that, in effect, saw him everywhere, in the life style and behavior of the greedy. By the 1690s, however, Hobbes had few, if any, public defenders in a philosophical or intellectual sense. The Newtonians were most energetic and innovative, therefore, in devising arguments against the new generation of freethinkers, figures such as John Toland (1670–1722) and his friends. They grew more menacing than Hobbism simply because they were alive and well, devising assaults against the church, and at the same time being active in the political arena.

In consequence of this new challenge epithets like deist, atheist, libertine, and finally pantheist were bandied about so much that they lost all useful meaning. It is preferable to call the church's radical opponents freethinkers, a term sufficiently loose, yet part of contemporary parlance,[5] to allow for the divergent views found among the intellectual and political non-conformists. One of their number, William Stephens (1647?–1718), a friend to John Toland and Anthony Ashley Cooper, third earl of Shaftesbury, tried to account for the "growth of deism." With considerable perception he laid blame on the

[4] Much of this antichurch sentiment, interestingly enough, came from high-churchmen. See William J. Cameron, ed., *Poems on Affairs of State. Augustan Satirical Verse, 1660–1714*, V, *1688–97* (New Haven, 1971), 309 passim. Cf. Harley 1604, f. 51, et seq., Sloane 1024, Harley 7315, ff. 146–147, B.L.

[5] The term first appears in a tract by one Sebastian Smith, *The Religious Impostor: or the Life of Alexander, A Sham-Prophet Doctor and Fortune-Teller. Out of Lucian. Dedi. to Doctor S—lm—n, and the Rest of the Free-Thinkers, near Leather Sellers-Hall* (London, 1692). Here freethinking is equated with loose, Grub Street living on the part of a small, prophetic sect.

Revolution and its consequences, and his comments deserve
lengthy quotation:

Now the oldest *Deists* of my Acquaintance having conceiv'd so
great a Prejudice against the Christian Faith, from the Behaviour
of the Clergy, and having levened their Disciples therewith, it
has fal'n out unhappily, that the late Revolution has by another
way also confirmed them in this their Prejudice.

For the late happy Revolution, (which came on too soon, and
was cut off too short) though it was not so highly beneficial to
us, as was by some expected, was yet of very great Importance.
But as there is nothing in this World ever so good, but what hath
some appending disadvantage; so by meer Accident this Revolution, which has saved not only the Church of *England*, but (as I
hope) the whole Protestant Interest throughout the World, has
wonderfully encreased Mens Prejudices against the *Clergy*, and
so by false Consequence (such as Men through Resentment will
make) against the Truth of Religion it self. The old *Deists* tell
those of their Pupils, who never travelled abroad, that there is
now no need of going over the Water to discover that the name
Church signifieth only a *Self-interested Party*, and that the *Clergy*
have no Godliness but Gain. Have you not (say they) for many
Years together heard them Preach up the *Divine Right*, and indefeizable Authority of Kings, together with *Passive Obedience*, as
the chief distinguishing Doctrines whereby their Church approved
it self Apostolick beyond all other Churches? Nay, were not the
Doctrines of *Loyalty to the King*, insisted upon more than *Faith
in Christ?* and yet when their particular Interest required it, their
Doctrine of *Non-Resistance* was qualify'd by *Non-Assistance*, the
whole Stream of Loyalty was turn'd from the King to the Church,
the indefeizable Right was superseded by a miraculous Conquest
without Blood, the Oath of Allegiance to the *Divinely Rightful
King James* has its force allay'd by another Oath of the same Importance made to the *de facto* King *William* and Queen *Mary*,
and all this is Sanctify'd by the name of the Church, *i.e.* their own
Party and Interest, for the sake whereof it is done.[6]

[6] Stephens, *An Account of the Growth of Deism in England* (London, 1696), 10–11. The Revolution is also blamed in Anthony Hor-

Stephens know how to attack his adversaries squarely and effectively. He focuses on the church's ambivalence toward the Revolution, its embarrassing abandonment of divine right and passive obedience, and its subsequent pragmatic—Stephens calls it self-interested—accommodation to the constitutional settlement. The intellectual force and conviction in deism derived in large measure from political and ideological opposition to the clergy's power in government and society. Stephens notes that the Revolution came too soon and was too short. The freethinking coterie to which Stephens belonged and around which our discussion will center believed in a radical Whiggery and derived its political ideology largely from the Commonwealth tradition. One church contemporary put the matter succinctly: "I find the Republicarians in these parts sedulous to promote atheisme, to which end they spread themselves in coffyhouses and talk violently for it."[7]

Stephens' comments, although telling and perceptive, were hardly unbiased. Although an ordained churchman and rector at Sutton, Surrey, Stephens abandoned his political and intellectual commitment to the church for reasons that are not altogether clear. The tone of his writing displays disdain for the church's hierarchy and for what Stephens regards as their self-interested and Tory politics. By 1699 he had taken the freethinker, John Toland, into his home, and in 1706 he was hauled before the courts, fined, and pilloried, for having written against the government and for withholding evidence against one Thomas Rawlins, also suspected of having written against the government. Stephens was on reasonably close terms with the third earl of Shaftesbury and may even have received payments from him, as did Toland.[8]

neck, *An Antidote against a Careless Indifferency in Matters of Religion* (London, 1694), introduction.

    [7] *Letters of Humphrey Prideaux*, 162.

    [8] Charlett to Kennett; Ballard 7, f. 39, Bodleian; P.R.O. 30/24/20, f. 111, and f. 115; *DNB*, Rawlins was accused of authoring *A Letter to the Author of the Memorial of the State of England* (really by Stephens); P.R.O. 30/24/20, f. 111.

Although Stephens locates deism as a post-Revolution phenomenon, and in terms of number of adherents he was probably right, nonetheless its roots stretched well back into the seventeenth century. For instance, the belief that matter possessed an inherent life and motion was common to some radical sects during the mid-century Revolution.[9] Earlier still, Herbert of Cherbury (1583–1648) in his *De veritate* (1624) articulated a rational version of Christianity closer to the ideals of freethinkers like Anthony Collins and Matthew Tindal than to the natural religion of the latitudinarians. In the hands of the unorthodox almost any philosophical system could be made ungodly, and in that sense the Cartesian version of the mechanical philosophy contributed to the growth of freethinking both in England and on the Continent.[10] Perhaps the most unexpected connection I intend to explore between early eighteenth-century freethinking and previous unorthodoxies appears in the use to which John Toland was able to put the writings of Giordano Bruno. But my intention at this point is not primarily to explore the antecedents of freethinking and therefore the origins of the Enlightenment; that subject will require another book. Rather, in order to enhance our historical understanding of the social meaning of Newtonianism, we must analyze the composition and activities of the church's opposition.

In the early 1690s the threat of widespread atheism appeared more real than it had at any time after 1660 and to affect every level of the society. Treatises by divines urging the faithful

[9] Christopher Hill, *The World Turned Upside Down* (London, 1972), 112, 114, 176.

[10] A. Vartanian, *Diderot and Descartes: A Study of Scientific Materialism in the Enlightenment* (Princeton, 1953); Toland is the author of objections to Bayle's refutation of Dicearchus (226). See C. Gerhardt, *Die Philosophischen Schriften von Gottfried W. Leibniz* (Berlin, 1882), III, 68. H. Kirkinen, *Les Origines de la conception moderne de l'homme-machine: Le Problème de l'âme en France à la fin du règne de Louis XIV (1670–1715)*, in *Annales academiae scientiarum Fennicae* (Helsinki, 1960).

not to succumb poured from the presses. Bentley in his Boyle lectures spoke to the prosperous of London, but his arguments and others of a less sophisticated nature appear in tracts aimed at the literate, but humble and uneducated, craftsman or shopkeeper. Even Locke and Descartes were enlisted and simplified in the effort to prove that matter cannot think or that the soul is separate from the body.[11] For those whose future is uncertain, whose prosperity is tenuous, the argument from design should, it was argued, provide security. Order in the world comes only from a belief in God; without God men would be wild beasts and insecurity would turn to anarchy.[12] Freethinkers are dangerous because they believe "there's no Inferiors, But all were born upon the Level," and to make matters worse most of them are robbers and thieves.[13] Fantasies reached such heights that even the affluent third earl of Shaftesbury could be presumed to engage, at least by association, in such criminal practices.[14] By 1694, William Talbot saw atheism as rampant, proclaimed that without God men would go about unrestrained, and asserted that those who acknowledge nothing but matter and motion "labour to make a Party."[15]

The charge that the freethinkers formed a cabal or party occurs consistently in their opponents' literature. The historian is tempted to dismiss it entirely as a piece of official paranoia, but that would be unwise. Sufficient evidence exists, most of it unpublished, to posit that many of the freethinkers knew one another, socialized together, engineered literary projects, and even traveled about incognito[16] in London and then on the

[11] Richard Sault, *The Second Spira: Being a Fearful Example of an Atheist* (London, 1693).

[12] Clement Ellis, *The Folly of Atheism, Demonstrated, to the Capacity of the Most Unlearned Reader* (London, 1692), passim.

[13] Anon., *Freethinkers. A Poem in Dialogue. As Atheism Is in All Respects Hatefull* . . . (London, 1711), copy at Widener Library, Harvard.

[14] Ibid., 17.

[15] William Talbot, *Twelve Sermons Preached on Several Subjects and Occasions* (London, 1725), 4.

[16] "I lodg at M'r Ridgley's in le la Haye's Stre. It has a Door into the

Continent. Indeed, later in this chapter strong evidence will be presented to support the claim that John Toland belonged to a secret society from as early as the 1690s which can best be described, for lack of a better term, as an early Masonic lodge. Yet the freethinkers were never as well organized or as conspiratorial as the church liked to think—paranoia did occasionally grip even the calmest churchmen. Freethinkers were, however, in communication with one another, and they did on many occasions resort to secrecy. In a society where prosecution for disagreeable publications was still very possible[17] we should hardly be surprised that they preferred a private, and even secret, lifestyle. Private clubs were widespread in the Augustan period and freethinkers, like other people, resorted to them.

The techniques of spying were employed by both sides. There were paid informants who spied on the presses, generally in search of the purveyors of what was considered to be pornography.[18] Toland tampered with letters exchanged by churchmen, and letters sent to him turn up in the library of Lambeth Palace, the home Archbishop Thomas Tenison, with the inscription "Letter to Toland seized in Ireland" written on them.[19] Churchmen often met in private to discuss the latest perfidious book published by a freethinker. Meetings were held at Wake's home where Matthew Tindal's *Rights of the Christian Church* (1706) was discussed,[20] and when the dinner conversation of divines and university men turned to

---

[?], by which way one may come to me incognito." Toland to Shaftesbury, 1705, P.R.O. 30/24/20, f. 105.

[17] A survey of prosecutions can be found in Charles R. Gillett, *Burned Books: Neglected Chapters in British History and Literature* (New York, 1932).

[18] David Foxon, *Libertine Literature in England, 1660–1745* (London, 1964).

[19] Kennett to Charlett, 1701, Ballard 7, f. 48, Bodleian; MS 933, f. 55, Lambeth Palace Library, from "Aristodemon," obviously a code name.

[20] MS 1770, Aug. 14, 1706, Lambeth Palace Library.

"bookish matters" they would "rail at several pieces that had
of late been writ . . . [by] Toland, Tindal, Collins, Asgil,
[and] the Author of ye Tale of a Tub." Shaftesbury gleefully
reported one such occasion when his own *Letter concerning
Enthusiasm* (1708), published anonymously, was maligned in
his presence by churchmen who had no idea as to the author's
true identity. It was believed that the *Letter* had dealt cavalierly
with the prophecies.[21]

The freethinkers knew who their enemies were. In a manuscript sent to the Continent, Toland named his prime antag-
onists and included Samuel Clarke and Stephen Nye.[22] Clarke,
of course, had mentioned Toland by name in his Boyle lectures, and the antagonism between the freethinkers and the
Newtonians stands as one of the main themes in the intellectual
history of the early eighteenth century. Substantial evidence
exists to warrant the conclusion that Newton himself contemplated a direct answer to Toland's attack on his philosophy
which he intended to include in the 1706 edition of the
Opticks. I shall return to that evidence shortly. Toland's other
antagonist, Stephen Nye, was a Socinian of sorts, but he won
favor with Tenison for his attacks on Toland in a project
which Tenison may have even suggested and encouraged.[23]

The church feared the renewed energy of the freethinkers
not simply because they promoted every unorthodoxy from
deism to pantheism, but because they coupled freethinking
with radical republican politics. They actively sought to strip
the church of its political and social power and to enhance the

[21] P.R.O. 30/24/22/4, f. 68, 97. Cf. MSS ADD 4292, f. 63, B.L.

[22] Toland MSS, 10325, f. 134, National Library, Vienna. I fail to see
that these manuscripts add very substantially to the material already
available in English archives. But the knowledge of their existence is
both useful and valuable. See Franco Venturi, *Utopia and Reform in
the Enlightenment* (Cambridge, 1971), 60, and G. Ricuperati, "Liber-
tinismo e deismo a Vienna: Spinoza, Toland e il *Triregno*," in *Rivista
storica italiana*, 2 (1967), 628 ff.

[23] Nye to Tenison, Dec. 6, 1699, and Nov. 4, 1703, MS 953, Lambeth
Palace Library.

power of Parliament at the expense of king and court. The free-thinking movement of the early eighteenth century repre-sented in one sense a revival and continuation of traditional country opposition to the court. Both Robert Molesworth and Anthony Collins were gentlemen of lesser stature, and al-though they may have disagreed with the religious and philos-ophical ideas of Toland, they supported him financially and took him into their homes.[24] Although the church had never countenanced freethinking in any form, the Epicureanism found in court circles during the Restoration[25] posed no real threat to the church's political power when compared to the political ideals and activities of the post-1689 freethinkers. Churchmen failed to see that the leaders of the movement were unable and unwilling to seek mass support. Whereas the first earl of Shaftesbury had willingly courted the crowds in his bid for power, his grandson, the third earl, and his friends circulated in their own private world, wrote lengthy and ab-stract treatises, and provided no evidence as yet discovered to indicate that they sought to bring the poor or the socially disaffected into their movement. No better example of the political remoteness of the freethinking movement from the general populace can be found than Anthony Collins' activities as a county justice and treasurer in Essex.[26]

The freethinking radicals of the early 1700s never aroused a populace which, if it had known the depths of their irreligion, would undoubtedly have repudiated them. One prosperous

[24] MSS ADD 4465, ff. 22, 50, B.L. For Molesworth and Tindal see Historical Manuscripts Commission, *Report on Manuscripts in Various Collections*, III (London, 1913), 258. Toland's philosophizing after 1697 may have had something to do with the country vs. city debate of the early eighteenth century. Yet he is, at the height of his political influence and intellectual power, very much an urban and cosmopoli-tan figure.

[25] See Thomas F. Mayo, *Epicurus in England (1650-1725)* (Austin, Tex., 1934).

[26] James O'Higgins, S.J., *Anthony Collins, the Man and His Works* (The Hague, 1970), chap. 8.

businessman and pious Socinian, Thomas Firmin, did attempt, however, to spread his unorthodox version of Christianity to the gentry and also to "the Tradesmen whom he deals with to whom he seldom sends a pack of goods without a bundle of those Hereticall Pamphlets enclosed."[27]

The ultimate failure of the political goals of the freethinking movement should not obscure the danger it posed to the church at the time or obviate the importance of the intellectual alternative it posed to Newtonianism. Many of the freethinkers' ideas came to prominence in the Enlightenment, as radical undercurrents or as part of the fully developed and comprehensive atheism of philosophers such as d'Holbach and Diderot.[28] These ideas entered the Enlightenment on both sides of the channel through the writings of Toland, Collins, Tindal, and others, but also through their private, and sometimes secret, propagandizing efforts. The freethinkers did form a loose cabal, and there is no better source of information about its composition and activities than the information available to us about the most notorious freethinker of all, John Toland.

Born in Ireland of Catholic, and probably poor, parents, but converting to Protestantism sometime during his adolescence, Toland first turns up in Scotland as a university student in Glasgow and then at the University of Edinburgh where he received his M.A. in 1690.[29] His early years remain a total

---

[27] Prideaux to Tenison, Oct. 2, 1696, Gibson MSS 930, f. 56, Lambeth Palace Library.

[28] Lester Crocker, "John Toland et le matérialisme de Diderot," *Revue d'histoire litteraire de la France*, 52 (1953), 289–295; in 1768 the Baron d'Holbach translated *Letters to Serena* into French (previous manuscript translations did exist, MS 10325, National Library, Vienna), and his natural philosophy in *Système de la nature* (1770) bears striking resemblance to Toland's. Cf. L. Flam, "De Toland è d'Holbach," *Tijdschrift voor de studie van Verlichting* (Bruxelles), 1 (1973), 33–54. Toland's ideas were still alive in England in 1731; see Daniel Waterland, *Works* (Oxford, 1843), V, 42 passim.

[29] The best biography of Toland remains P. Desmaizeaux's introduction to *A Collection of Several Pieces by Mr. John Toland* (London, 1726) and 2d ed. (London, 1747). See also "An Abstract of the

mystery. We know that at Edinburgh he was tutored by
David Gregory, among others, and that Gregory made his
pupils perform exercises in Newtonian physics before they
took their degrees.[30] In all probability, Toland was exposed to
Newton's thought as part of his early education.

When Toland made his appearance in London in late 1691
or early 1692 he was a Presbyterian and a member of Daniel
Williams' congregation. In later years Toland would protest to
various churchmen that he was a loyal son of the church,[31]
but all evidence seems to confirm that he was simply telling
them what he thought they wanted to hear or what he thought
would annoy them. Interesting in juxtaposition to Toland's
subsequent anticlerical writings, both published and unpublished, is his first published piece, which was a defense of
Williams' ill-fated attempt to form a union of Presbyterians
and Congregationalists called the United Brethren. Toland
published his defense in Holland, in Le Clerc's *Bibliothèque
universelle et historique*.[32] Indeed this notorious freethinker

---

Life of the Author" prefixed to Toland's *A Critical History of the
Celtic Religion* (London, 1740); F. H. Heinemann, "John Toland and
the Age of Reason," *Archiv für Philosophie*, 4 (1950–1952), 35–66; H.
F. Nicholl, "John Toland: Religion without Mystery," *Hermathena*,
100 (1965), 54–65. For a bibliography of Toland's writings see W.
Deinemann, "A Bibliography of John Toland" (Diploma in Librarian-
ship thesis, University of London, 1953). I disagree with the account
found in G. R. Cragg, *From Puritanism to the Age of Reason: A
Study of Changes in Religious Thought within the Church of England, 1660–1700* (Cambridge, 1966, reprint), 136 ff. Of great value is
Venturi, 49ff; cf. my "John Toland and the Newtonian Ideology,"
Journal of the Warburg and Courtauld Institutes, 32 (1969), 307–331,
in which some of the topics discussed in this chapter first appeared. *A
Catalogue of the Graduates in the Faculties of Arts, Divinity and
Law, of the University of Edinburgh, since Its Foundation* (Edin-
burgh, 1858), 137–138, lists Toland as receiving his B.A. in 1690.

[30] Desmaizeaux, ed., *Toland*, vii–viii; Alexander Bower, *History of
the University of Edinburgh* (Edinburgh, 1817), I, 314.

[31] Ballard MSS, V, f. 33 in 1694, Rawlinson, MS c. 146, in 1706, Bodleian; in 1707 Gibson MS 930, f. 229, Lambeth Palace Library; and
Thomas Sharp, *The Life of John Sharp* (London, 1825), I, 273–275.

[32] I have accepted this letter to Le Clerc as Toland's first published

began his public career as a student for the Presbyterian ministry; Williams' group sent him to Holland to study theology and even gave him a stipend.[33]

From their point of view it must have been money ill-spent. Toland went to Holland in late 1692 or early 1693 and returned to England in August of that year a changed man. If we presume that Toland left England as a committed Christian and Presbyterian—and there seems no reason to doubt it—then his experiences in Holland produced in him an intellectual revolution. As far as we know he went to Utrecht and probably attended a few classes at the university there as well as at Leiden. For a time he resided near Amsterdam,[34] but the center of his intellectual development appears to have been the home of the Quaker refugee and friend of John Locke, Benjamin Furly. Furly's salon possessed one of the best libraries of rare and heretical works in Holland at that time, and his visitors included Locke and, later, in Toland's time, Jean Le Clerc, the Arminian theologian, F. M. van Helmont, the Hermetic philosopher and alchemist, and in the early eighteenth century, the third earl of Shaftesbury.[35] In this

---

work, despite his claim in *Christianity Not Mysterious* (London, 1696), xxiv, to have written a tract entitled *Systems of Divinity Exploded,* which was never, as far as I can see, published. Toland has sometimes been confused with John Tutchin, the real author of the *Tribe of Levi* (1691). For the latter see *Bibliothèque universelle et historique* (2d ed., Amsterdam, 1699), XXIII, 505–509.

[33] Toland received £16. Alexander Gordon, *Freedom after Ejection: A Review (1690–92) of Presbyterian and Congregational Nonconformity in England and Wales* (Manchester, 1917), 182.

[34] Ibid., 182. The *DNB* is wrong in stating that he spent two years at the University of Leiden. G. Bonno, "Lettres inédites de Le Clerc à Locke," *University of California Publications in Modern Philology,* 52 (1959), 69. Cf. F. H. Heinemann, "John Toland, France, Holland and Dr. Williams," *Review of English Studies,* 25 (1949), 347.

[35] William Hull, *Benjamin Furly and Quakerism in Rotterdam* (Philadelphia, 1941), 137. See also *Bibliotheca Furliana* (Rotterdam, 1714). See MSS ADD 4283, ff. 265–266, Dec. 20, 1700, B.L., for van Helmont's influence on Furly. Furly's correspondence is located amid the Shaftesbury MSS at the P.R.O. Part of it has been published in

setting the young Toland would have been exposed to the liberal and antitrinitarian theology of Le Clerc and Philip van Limborch as well as to Furly's own free-spirited and republican version of sectarian Protestantism. Locke's ideas were certainly discussed, and Le Clerc's interest in English thought included an acquaintance with Newton's ideas.[36]

Toland returned to England not as a Presbyterian minister but as a freethinker. Furly wrote enthusiastically to Locke about him:

I find him a freespirited ingenious man, that quitted the Papcy in James' time when all men of no principle were looking toward it, and having now cast off the yoke of Spiritual authority, that great bugbear, and bane of ingenuity, he could never be persuaded to lower his neck to that yoke again, by whomsoever claymed; this has rendered it somewhat difficult to him to find a way of substance in the world.[37]

Through these Dutch acquaintances Toland met Locke, for whom he brought letters from Le Clerc. Furly asked Locke to place Toland as a tutor and even indicated that Toland knew Locke to be the true author of the *Essay on Toleration*. Le Clerc also wrote favorably about Toland to Locke.[38]

Perhaps upon their first meeting Locke sensed something odd and even dangerous about Toland. He never aided him financially, and Toland went to Oxford where he immediately fell in with a circle of "ingenious men" to whom he was introduced by Thomas Creech, the translator of Lucretius.[39] Locke also had acquaintances in that circle, among them John Freke who wrote to Locke about Toland's comings and goings in

---

Benjamin Rand, ed., *The Life, Unpublished Letters, and Philosophical Regimen of Anthony, Earl of Shaftesbury* (London, 1900).

[36] Cf. A. Barnes, *Jean Le Clerc et la republique des lettres* (Paris, 1937); Bonno, 16.

[37] Heinemann, "Toland, France," 348. Cf. Bonno, 67–69.

[38] Sept. 11, 1693, Bonno, 69.

[39] Desmaizeaux, ed., *Toland*, II, 293; letter dated January 1694 (O.S.).

Oxford.[40] Those letters proved to be significant for Locke's intellectual development and his writing of the *Reasonableness of Christianity* (1695).

In his discussion of Toland and Locke, Leslie Stephen erroneously claimed that Toland wrote his first major work, *Christianity Not Mysterious* (1696), simply as an attempt to gain notoriety by boasting of intimacy with Locke and by engrafting his speculations onto Locke's version of Christianity revealed in the *Reasonableness*.[41] Stephen did not know that Toland embarked on writing *Christianity Not Mysterious* at least nine months before Locke began his own treatise[42] and that the publication of Toland's attack on Christianity in 1696, one year after Locke's defense, was probably determined by the lapsing of the Licensing Act. But the relationship between the *Reasonableness* and *Christianity Not Mysterious* does not end there. We now know that Locke had a copy of Toland's manuscript before it was published and precisely at the time when Locke embarked upon his defense of Christianity according to his own principles.[43] It was Locke who could be said to have answered Toland in the *Reasonableness* and not vice versa.[44]

[40] March 29 and April 9, 1695, Locke MS c. 8, ff. 193–194, Bodleian. Cf. John C. Biddle, "John Locke on Christianity: His Context and His Text" (Ph.D. diss., Stanford University, 1972), 17n.

[41] Stephen, *History of English Thought in the Eighteenth Century* (New York, 1962, reprint ed.), I, 78.

[42] Desmaizeaux, ed., *Toland*, II, 312, letter from Oxford, May 30, 1694, edited to conceal the signature. Cf. S. G. Hefelbower, *The Relation of John Locke to English Deism* (Chicago, 1918), 158–159.

[43] John Freke to Locke, April 9, 1695, MS Locke, c. 8, f. 194, Bodleian: "Deepest thanks for your of the 8th which I received together with Mr T's Papers but give me leave of you that I hoped you would have said something to me of your opinion both of his Tract (I mean as much as you have seen of it) and of the man with respect to the Resolutions he seems by his letters to have taken for my own part I confess I have no great satisfaction in either." Freke seems to write also on behalf of Edward Clarke.

[44] In my "John Toland and the Newtonian Ideology," 311, I noted that Toland had begun writing his tract in 1694. Subsequently during

How much of Toland's completed version of *Christianity Not Mysterious* Locke had actually read in the spring of 1695 we cannot know. But he certainly would have read enough to discover that Toland was using his epistemological principles from the *Essay Concerning Human Understanding* (1690) to argue that "whoever reveals anything his words must be intelligible . . . and the matter possible. This Rule holds good, let God or Man be the Revealer."[45] Of course, Toland was twisting Locke's epistemology to suit his own ends, but the twists were subtle and Toland's use of Locke was both intelligent and perceptive.[46] Essentially Toland took Locke's statement in the *Essay* that "reason is the proper judge; and revelation, though it may, consenting with, confirm its dictates, yet cannot in such cases invalidate its decrees"[47] and argued that the data of revelation must conform to the ideas formed by the reasoning mind derived from the material presented by the senses. Therefore, if the so-called truths of revealed religion contradict human reason they are simply untrue.

In the *Reasonableness*, Locke argued persuasively that the truths of revelation are of necessity reasonable. By 1696, however, the damage had been done. Locke's views had been associated with those of the freethinkers, Toland was so taken with his abilities that he went to Ireland boasting that he was a friend of Locke and Le Clerc, and liberal Christians such as

the academic year of 1970–1971 I met with John Biddle who was completing a doctoral dissertation on Locke in Cambridge. In the course of our discussions I hypothesized, without evidence or great conviction, that Locke might have written his tract in answer to Toland. Biddle discovered the crucial piece of evidence, buried in Locke's unpublished correspondence, that completely reverses our understanding of the Locke-Toland relationship. Credit for this reversal rightfully should go to him.

[45] Toland, *Christianity*, 47.
[46] See my "John Toland and the Newtonian Ideology," 311–312, and John Yolton, *John Locke and the Way of Ideas* (Oxford, 1956), 118–126.
[47] Locke, *Essay Concerning Human Understanding*, ed. A. Fraser (New York, 1959), II, 423.

Le Clerc, Limborch and the Socinian, Thomas Firmin, dis-
avowed any connection, either personal or intellectual, with
Toland.[48] The whole episode between Locke and Toland
heightened the church's suspicion of the former[49] and provided Toland with considerable fame and notoriety. It also
illustrates how easily the foundations of latitudinarian Christianity could be seriously attacked.[50]

For his accomplishment Toland earned universal and public
condemnation from churchmen. In Ireland he was attacked
from the pulpits, in England pamphlets poured from the
presses and the Middlesex Grand Jury ordered that *Christianity Not Mysterious* be burnt. The lower house of convocation wanted to prosecute the author and his book, but
Archbishop Tenison informed them that they no longer
possessed the legal right to do so.[51] The frustration of the
lower house's efforts to control heresy, restrained by the
church's hierarchy, exacerbated relations between the ecclesiastical authorities and the lower clergy. By the reign of Anne
the Tory lower house of convocation made it clear to the
Whiggish upper house that it believed the church to be in
danger and that the church's leadership should be held accountable for the deplorable spread of heresy and irreligion.
Undeterred by questions of legality, the lower house once
again attempted to prosecute Toland in 1704, but its efforts
only met with frustration.[52]

[48] P. Desmaizeaux, ed., *The Works of John Locke* (London, 1759),
III, 538, 616; cf. H. Fox Bourne, *The Life of John Locke* (London,
1876), II, 416.

[49] Desmaizeaux, ed., *Locke*, 537–538.

[50] The latitudinarians were worried; see Edward Stillingfleet, *A Sermon Preached before the King and Queen . . . on Christmas Day,
1693* (London, 1694), 6–7; also T. Becconsall, *The Grounds and
Foundation of Natural Religion . . . in Opposition to . . . Modern
Scepticks and Latitudinarians* (London, 1698).

[51] E. Carpenter, *Thomas Tenison, Archbishop of Canterbury* (London, 1948), 82. Cf. G. V. Bennett, *The Tory Crisis in Church and
State 1688–1730* (Oxford, 1975), 58–59.

[52] "Observations made by the President and his Suffragen Bishops in
the Convocation of Canterbury upon a paper delivered . . . by the

As early as 1694 when Toland was in Oxford his activities drew the attention of churchmen. Edmund Gibson (1669–1748), a fellow of Queens, wrote derogatorily to a friend about him and claimed that Toland had associated with Rosicrucians in Edinburgh.[53] That bizarre accusation was almost certainly a fiction especially because we have good reason to doubt that the Rosicrucians ever existed as an organized society or party.[54] Yet it is significant that of all the possible affiliations Toland might have had, Gibson tries to label him as a Rosicrucian, a member of the Brethren of the Rosy Cross.

This mythical fraternity supposedly sprang to life in Germany sometime around 1615, inspired by the Hermetic philosophizing of John Dee as adopted by the religious mystic and reformer Johann Valentine Andreae (b. 1586). Although an organized Rosicrucian fraternity almost certainly never existed, the movement for Enlightenment proclaimed most coherently in the writings of Andreae stirred the imaginations of countless European reformers, not least of them Descartes.[55] The followers of Rosicrucianism wanted a universal and general reformation of the whole world, instituted through the philosophies of magic and cabala, whereby antichrist would eventually perish and true science and learning, the ancient and pristine wisdom, would triumph in a new and golden age. We must ask whatever inspired Gibson to imagine that Toland could have subscribed to such an essentially Christian and Platonic ideology.

By the late seventeenth century, and of course earlier, the Rosicrucian phenomenon had come to suggest the formation of secret societies.[56] I suspect that if one did a history of the

---

prolocutors of the Lower House," 1704; also aimed at William Coward; Patrick MSS, Queens' College, Cambridge.

[53] Heinemann, "Toland and the Age of Reason," 40.

[54] Frances Yates, *The Rosicrucian Enlightenment* (London, 1972), chap. 3.

[55] Ibid., 207, 114–17.

[56] Ibid., 207. Cf. L. Wright, *The Soul the Body at the Last-Day, Proved from Holy Writ . . .* (London, 1707).

218 The Newtonians and the English Revolution

usage of the term "Rosicrucian" in the seventeenth century it
would emerge, in common parlance, as an epithet applied to
almost any social gathering of a suspicious and often secret
nature. In seventeenth-century England, especially at the time
of the civil wars and onward, there must have been many such
gatherings. Can it be only accidental that the first two clear
references we have to membership in a Masonic Lodge come
from the 1640s? In 1641, Sir Robert Moray was admitted to a
Masonic Lodge in Edinburgh and in 1646, Elias Ashmole
joined a lodge in Lancashire that accepted members of either
political conviction.[57] One serious scholar interested in the
history of Freemasonry—and the subject has been bedeviled by
much quackery—now acknowledges that "the European
phenomenon of Freemasonry almost certainly was connected
with the Rosicrucian movement."[58]

When Gibson accused Toland of being a Rosicrucian he
most probably meant that rumors had associated Toland in
Edinburgh with membership in a secret club or society, suspi-
cions which quite possibly Toland's social behavior in Oxford
tended to confirm. In all probability Edinburgh never sported
a single Rosicrucian. Nevertheless there is irony in Gibson's
accusation, for it can be demonstrated with some conviction
that from the 1690s (possibly from as early as his Edinburgh
days) until his death in 1722, Toland belonged to a secret
society that, largely for lack of a better term, I shall call
Masonic. In order to substantiate that claim it is necessary to
discuss briefly the accepted history of the origins of Freemasonry.[59]

[57] Yates, 210.
[58] Ibid., 218; cf. Yates, *Giordano Bruno and the Hermetic Tradition*
(London, 1964), 274, 414–416, 423, and her *The Art of Memory*
(London, 1966), 303–305.
[59] The most comprehensive standard history remains Dudley
Wright, *et al.*, eds., *Gould's History of Freemasonry throughout the
World*, originally by Robert F. Gould, (New York, 1936), to be
treated with some caution. Also very useful is D. Knoop and G. P.
Jones, *The Genesis of Freemasonry* (Manchester, 1947). For a useful
summary of important Masonic dates see Robert Amadou, *Fasta Lato-*

Leaving aside the claims of Masons themselves that their organization dates back to ancient Egyptian times, modern Freemasonry of a speculative nature emerged first in seventeenth-century England. Operative masonry, the lodges of practicing masons, began at that time to admit nonmasons. They took over the lore and ritual of the operative lodges— "the trade secrets of the operative masons became the esoteric secrets of the speculative masons"[60]—and certain intellectual ideals of the Hermetic tradition came to occupy a prominent place within Freemasonry. Religious toleration, free-spirited intellectual inquiry intended for the improvement of man's condition, and dedication to the search for esoteric wisdom about nature appear as goals in the first official Masonic publication, James Anderson's *The Constitutions of Freemasons* (London, 1723). By 1717 and the formation of the Grand Lodge of London, the first official and public proclamation of the existence of Freemasonry, operative masonry had begun to disappear, having been transformed into the speculative and secret society that became so commonplace in eighteenth-century Europe.

The history of the emergence of speculative Freemasonry from the time of Ashmole's membership in 1646 until the founding of the Grand Lodge in 1717 remains cloudy and uncertain. But there are clues to Toland's involvement in the early history of Freemasonry. In 1676 we know that certain "accepted Masons" dined with the Green Ribboned Cabal, a Whig club that was to play a prominent role in the Exclusion Crisis.[61] In 1686, Robert Plot declared the Masonic movement

---

morum ou annales maçonniques des origines à nos jours, Extrait du Tome XI des Travaux de Willard de Honnecourt (Paris, 1974).

[60] J. M. Roberts, *The Mythology of the Secret Societies* (London, 1972), 21. For French Freemasonry, pioneering work has been done by Pierre Chevallier, *Les Ducs sous l'Acacia ou les premiers pas de la Franc-maçonnerie française 1725–1743* (Paris, 1964). Cf. Harry Carr, *Lodge Mother Kilwinning, No. 0. A Study of the Earliest Minute Books* (London, 1961).

[61] Yates, *Enlightenment*, 217; Roberts, 59n.

to be widespread throughout the country.[62] In the 1690s the existence of two Masonic lodges, one in London, the other in York, appears to be accepted by serious historians of the movement. The official and first historian of the Grand Lodge, James Anderson, claimed that the Whig member of Parliament, Sir Robert Clayton, headed the London Lodge in the 1690s, and his claim seems to have been accepted. In York one of the only known members of that lodge was William Simpson.[63] Also in 1698 an anonymous pamphlet accused the Masons of being the antichrist and warned Londoners, "mingle not among this corrupt People lest you be found so at the World's Conflagration."[64]

By the time of the founding of the Grand Lodge in 1717 the London Freemasons were eminently respectable. Among prominent founders were Jean-Theophile Desaguliers, the Newtonian, John, second duke of Montagu, and John Beal— all fellows of the Royal Society. They commissioned James Anderson, a Scottish divine, to draw up a constitution for the Lodge and "charges" required of its members. The first charge or duty of the Freemason, as Anderson saw it, deserves full quotation: "A Mason is obliged by his Tenure to observe the Moral Law . . . and if he rightly understands the Craft, he will never be a *Stupid Atheist*, nor an *Irreligious Libertin*, nor act against Conscience."[65]

On the basis of the evidence available I would argue that Anderson and his fellow Masons were so anxious to prevent libertines and atheists from becoming members of the Grand

[62] Plot, *The Natural History of Staffordshire*, quoted in D. Knoop, G. P. Jones, D. Hamer, *Early Masonic Pamphlets* (Manchester, 1945), 31 et seq. Dorothy Schlegel has pointed out to me the Minerval symbolism on the frontispiece to Plot's 1686 edition.

[63] A. S. Frere, *Grand Lodge, 1717–1967* (Oxford, 1967), 32.

[64] Knoop, Jones, Hamer, 34–35.

[65] James Anderson, *The Constitutions of the Right Worshipped Fraternity of the Free and Accepted Masons* (London, 1756), 143. Italics added. Cf. Knoop, Jones, Hamer, 46, for Samber's claim that the Masons are not atheists.

Lodge precisely because some lodges were infected, perhaps even dominated, by freethinkers. The new Grand Lodge, an amalgam of a number of London lodges, was attempting to cleanse the Masonic movement of members whose views and activities offended the clerical representatives of orthodox Christianity. Among the confiscated letters to John Toland located in Lambeth Palace Library is a letter signed with a code name which states, "About a month ago they were all treated at Sr. R. Clayton's House at the City's charge. Sir Robert after dinner proposed to your Friend the recommendation of you to all of them."[66] However, the correspondent informs Toland that his bizarre and flamboyant conduct in Ireland has alienated his powerful and unnamed patron who spoke against him at the meeting. In another letter the members of Toland's circle in London are named: John Methuen, lord chancellor of Ireland, Mr. Freke (probably John Freke, Locke's informant on Toland), Mr. Clarke (whom I have not been able to identify),[67] Mr. Rawlins (probably Thomas Rawlins), and the letter is signed by William Simpson, possibly that known member of the York lodge. He adds, almost as an afterthought, "I had forgot to tell you, that you have hazarded your Credit with my Lord Spencer by owning the manuscript book of Servetus, which Mr. Firmin has since laid claim to on behalf of a friend of his."[68] The existence of this coterie was proclaimed in a ribald ballad which was sung on the streets of London against Methuen, Toland, Molesworth, and Walter Moyle (the last three Commonwealthmen).[69]

[66] MSS ADD 4295 f. 28, June 1697, B.L.; MS 933, f. 55, dated June 1, 1697, Lambeth Palace Library.
[67] He might have been Edward Clarke, Locke's great friend and correspondent.
[68] MS 933, f. 74, Lambeth Palace Library.
[69] A. D. Francis, *The Methuens and Portugal* (Cambridge, 1966), 69, 356–357. It is claimed that Toland was to have been Methuen's secretary in Ireland and that he "has lately set up a new sect of religion beyond the Socinians." Cf. for Clayton and Toland, F. H. Ellis, ed., *Poems on Affairs of State*, VI (New Haven, 1970), 409.

These seized letters made no mention of Freemasonry. Yet other evidence, to be discussed shortly, ties Toland to some sort of Masonic group. Is it not possible that the fellowship to which Toland belonged in the 1690s and which met at the home of Sir Robert Clayton and included one William Simpson was the lodge Clayton was reputed to head? All of the persons referred to in Simpson's letter were prominent Whigs, and it appears that Toland was employed by them, somewhat unsatisfactorily, as an emissary to his native Ireland. Among his activities abroad may have been the circulation of Socinian treatises, in this case a tract by Servetus.

During his trip to Prussia in 1702, Toland apparently claimed that he belonged to, or headed, a secret sect or society.[70] It was probably the same group with which Toland claims he discussed and analyzed Bruno's *Spaccio de la bestia trionfante* (London, 1584).[71] In 1712, Toland wrote Georg Wilhelm, baron of Hohendorf, an intimate friend of Eugene of Savoy, that his philosophical liturgy was near completion. If it bore any resemblance to the liturgical rites Toland published in 1720 in his *Pantheisticon* it was intended as part of the services of some secret sect.[72] According to Toland it met in England and on the Continent, and its members should be described as pantheists, a word Toland invented in 1705 to describe his own religious views. Toland's assertions in the *Pantheisticon*, as well as his pantheistic materialism, have always puzzled those historians who have taken Toland and his philosophy at all seriously.[73]

[70] Jean P. Erman, *Mémoirs pour servir a l'histoire de Sophie Charlotte, Reine de Prusse* (Berlin, 1801), 198.

[71] Jacques de Chaufepié, ed., *Nouveau dictionnaire historique et critique . . . de Mr. P. Bayle* (Amsterdam, 1750), II, 455; and *Nova Bibliotheca Lubecensis* (Lübeck, 1756), VII, 158. See also Desmaizeaux, ed., *Toland*, II, 376–378.

[72] MSS ADD 4295, f. 20, B.L.; *Pantheisticon: or the Form of Celebrating the Socratic Society* (1st pub. 1720, English trans., London, 1751), 14, 70–71.

[73] See Albert Lantoine, *Un Precurseur de la Franc-Maçonnerie,*

Certainly the church paid very close attention to Toland and his associates. In order to understand its concern we must attempt to determine if secret societies, with a freethinking membership, did in fact exist in this period. If it can be determined that some of these secret societies were discernibly Masonic, this knowledge would add immensely to our understanding of the social organization of the early Enlightenment. The evidence from Toland's manuscripts makes plausible his claim to membership in a secret society. In 1720 when Toland resided at the home of another important English freethinker, Anthony Collins, he wrote about his philosophy and liturgy to a confidant, Barnham Goode,[74] one of the Grub Street crowd immortalized by Pope in the *Dunciad*. Describing Collins to Goode as your "fellow collegian," Toland elaborated on the meaning of certain abstruse portions of the *Pantheisticon*. The letter contains no absolute proof that Toland, Collins, Goode, and others met together or used Toland's liturgy, but to conclude the plausibility of such meetings seems entirely reasonable.

Can Toland's society or group be termed Masonic? If the historian assumes that the official description of Freemasonry given by Anderson in 1723 must be the only true one, then obviously Toland's group does not qualify. But this would be to take Anderson at face value and to ignore the evidence we now possess. Before the formation of the Grand Lodge and its subsequent organizational efforts, Masonic lodges may have served many functions. Conviviality, male companionship

---

John Toland, 1670–1722, suivi de la traduction française du "Pantheisticon" (Paris, 1927).

[74] MSS ADD 4295, ff. 39–40, B.L. Also mentioned in the letter is one Mr. Ingram. Possibly Goode is the B. G. addressed in "Physic without Physicians" in Desmaizeaux, ed., *Toland*, II, 273ff. Could Mr. Ingram be Arthur Ingram, sixth viscount Irwin, admitted to a private lodge in Yorkshire in 1726? See Dudley Wright et al., eds., *Gould's History of Freemasonry* (London, 1931?), II, 241. The Ingram MSS at the Leeds Public Library, Sheepscar Branch, unfortunately do not help to answer this question.

without the necessity of conventional sexual role-playing, epicurean dining and drinking were important; but the element of secrecy so essential to Masonry could also provide freethinkers like Toland with the anonymity they required. Once nonmasons were admitted into lodges and Freemasonry assumed its speculative character, is it not reasonable to imagine that before the Grand Lodge of 1717 inserted order into the movement many kinds of lodges, serving many different functions, had come into being?

Amid Toland's unpublished papers exists yet another record of a secret society that appears discernibly Masonic[75] with which Toland, in some way, must have been associated. The "Chapitre General des Chevaliers de la Jubilation" met at the Hague in 1710, had a constitution and grand master, and was dedicated to good food and drink and to the avoidance of matrimony "le Tombeau des Ris et des Jeux."[76] Some or all of its membership signed this particular manuscript which records the minutes of one of the group's meetings: G. Fritsch was Grand Master, with M. Böhm, G. Gledisch, C. Levier, Bernard Picart, M. De Bey, and Prosper Marchand as members. Marchand and Böhm edited and published the 1720 edition of Bayle's *Dictionnaire;* Picart, one of the finest engravers of the eighteenth century, illustrated editions of classical authors, including the 1732 edition of Ovid's *Metamorphoses;* and Levier, Böhm, and Fritsch traded books and arranged for the printing of editions between England and Holland.[77]

[75] Cf. my "An Unpublished Record of a Masonic Lodge in England: 1710," *Zeitschrift für Religions-und Geistesgeschichte,* 22 (1970), 168–169. Cf., Roberts, 18n. I do not see what other term could possibly describe this group. Further research has shown that they did not meet in London.

[76] MSS ADD 4295, ff. 18–19, B.L. I know of only one other club that may have been similar, the Ordre des Chevaliers de la Joye. See M. A. Dinaux, *Les Societes badines, bachiques, littéraires et chantantes* (Paris, 1868), I, 421–424, and my *The Radical Enlightenment* (1981).

[77] Frank Manuel, *The Eighteenth Century Confronts the Gods* (Cambridge, Mass., 1959), 4, 7. My thanks to Mr. Marmoy, Huguenot

Marchand knew a great deal about clandestine manuscripts and had read Bruno's *Spaccio* and knew of Toland's connection with it.[78]

Marchand also had a hand in the Continental circulation of the *Pantheisticon*. Among his manuscripts at the University of Leiden are extensive notes on the *Pantheisticon*, which may have been intended for a new edition of it. In these notes meant for the use of Thomas Johnson, a bookseller at The Hague, Marchand claims to have little familiarity with Masonic lodges, but then points out the similarity between Toland's bacchanalian liturgy found in the *Pantheisticon* and the "singular and bizarre" ceremonies of the Masons.[79] That such an association should have been made by Marchand, himself a member of a secret society and in some way an acquaintance of Toland's, adds further evidence for the contention that there were freethinking lodges in existence, probably on both sides of the Channel, during the first quarter of the eighteenth century. One hypothesis that merits careful study suggests that this type of Freemasonry should be described as "Minerval," because the figure of Minerva recurs in the frontispiece of works by the third earl of Shaftesbury, among others.[80] That figure coupled with other symbols recurs frequently in the engravings of Bernard Picart, a member of the Knights of Jubilation who also did an interesting engraving for the 1720

Society Library, University of London, for assistance. Extensive information about their activities can be found in MSS ADD 4285, B.L., and in the Marchand MSS, University Library, Leiden.

[78] *Dictionnaire historique* (The Hague, 1758), I, 318–325.

[79] Marchand MSS 62, University Library, Leiden; see E. F. Kossman, "De Boekhandel te s'Gravenhage tot het eind van de 18ᵈᵉ eeuw," *Bijdragen tot de Geschiedenis van der Nederlandschen Boekhandel*, 13 (1937), 206–210, for information on Johnson, an Englishman with a strong interest in freethinking literature. Cf. Aubrey Rosenberg, *Tyssot de Patot and His Work, 1655–1738* (The Hague, 1972), 92–93; and Johnson to Desmaizeaux, MSS ADD 4284, ff. 177–191, B.L.

[80] Dorothy Schlegel, "Freemasonry and the *Encyclopédie* Reconsidered," *Studies on Voltaire and the Eighteenth Century*, 90 (1972), 1433–1460.

edition of Bayle's *Dictionnaire*. There is a story yet to be told about these clandestine groups and their important role in the development and spread of early Enlightenment culture.

Other members of Marchand's circle appear to have been interested in clandestine manuscripts. Böhm circulated the infamous *De tribus impostoribus* (Rotterdam, 1721),[81] which labeled Christ, among others, as an impostor. All the members of the Knights of Jubilation were Protestant refugees, they were friends of Pierre Desmaizeaux, the editor of Toland's collected works, and they probably were familiar with Benjamin Furly. Their mutual correspondence was lively, and some of it is preserved at the University Library, Leiden. Letters are signed "Le Grand Maistre" and reference is made to the order and its membership. This group was part of an international society, centered in England and Holland, that was freethinking in its sentiments and generally republican in its politics.

Into these freethinking circles whose existence so plagued the church in the early eighteenth century we can now with some certainty place the Knights of Jubilation, John Toland, Anthony Collins, Robert Clayton and his friends, John Trenchard, William Stephens, Benjamin Furly, Anthony Ashley Cooper, third earl of Shaftesbury, Robert Molesworth, Matthew Tindal, John Asgill, and some lesser lights who associated with Toland and Collins at the Grecian Coffee House and read with them Bruno's *Spaccio*.[82] They were not without patrons or friends in high places. Toland attracted the attentions of powerful Protestant leaders on the Continent, Prince Eugene of Savoy and John William, the elector of the

---

[81] J. S. Spink, "La Diffusion des idées matérialistes et antireligieuses au début du XVIII siècle: Le Theophrastus redivivus," *Revue d'histoire littéraire de la France,* 44 (1937), 254n.

[82] Shaftesbury MSS, P.R.O. 30/24/22; 30/24/28/8. Stephens to Toland, May 1717, MSS ADD 4465, f. 12, B.L.; MSS ADD 4282, f. 192; ADD 4465, f. 27, B.L.; T. Hearne, *Remarks and Collections* (Oxford, 1906), III, 202; IV, 172. Cf. O'Higgins, 12, 77.

Palatinate, who praised and recommended him.[83] Lord Aylmer (d. 1720), a leading military figure and Whig member of Parliament, possessed a copy of an anticlerical and freethinking piece by Toland entitled "Bruno's Sermon."[84] Shaftesbury, although not always pleased by Toland's behavior and rejecting totally his pantheistic materialism, gave him an annual stipend.[85] And Robert Harley was able to employ him as a pamphleteer representing country sentiment.[86] In 1711, Whiston and Samuel Clarke held meetings with Tindal and Collins in an effort to convince them of the truth of the Christian religion. Their efforts met with failure, and the church remained convinced that atheism, of every type and description, infected the Whig party.[87]

Toland's freethinking circle had succeeded in merging the republicanism of the English Revolution with religious and philosophical radicalism. Specifically in the case of Toland, and particularly in the years after 1696, he combined republicanism with what I shall call his pantheistic materialism,[88] and most important, he culled this radical philosophy of nature from the writings of Giordano Bruno.

Bruno, the supreme Renaissance magus, had preached an esoteric, court-centered philosophy as an alternative to either the Catholic or Protestant Reformations. Toland transformed it into a philosophical justification for republican politics. His

[83] Venturi, 60. Cf. O'Higgins, 77. MSS ADD 4295, f. 15, B.L. For an account of Toland in Vienna see Powell to Isham, Berlin, 1708, Isham MSS 4636, Huntingdon Record Office, Huntingdonshire.

[84] MSS ADD 4295, f. 43, B.L.

[85] P.R.O. 30/24/21, f. 225.

[86] P.R.O. 30/24/21, f. 237. Cf. A. O. Aldridge, "Shaftesbury and the Deist Manifesto," *Transactions of the American Philosophical Society*, 41, pt. 2 (1951). Shaftesbury awaits a comprehensive study. On Toland and Harley, Angus McInnes, *Robert Harley, Puritan Politician* (London, 1970), 83–84.

[87] O'Higgins, 77; G. Holmes, *British Politics in the Age of Anne* (London, 1967), 57.

[88] See John Witty, *The First Principles of Modern Deism Confuted* (London, 1707), iv, xvi.

place in the English republican tradition and his comings and goings throughout Europe during the early 1700s in support of Protestant republicanism have been intelligently discussed by Franco Venturi.[89] What must be emphasized is the inspiration Toland received for his activities from his reading of Bruno, whose works were constantly at his side during many of his forays onto the Continent. Indeed over a century after Bruno's death in 1600 how much Toland resembles him in spirit as he wanders about the courts of Europe teaching what he has culled from Bruno's *Spaccio*. When we remember Toland's love of secrecy and his involvement in what was probably a version of Freemasonry, and the origins of that odd movement in the Hermetic tradition, are we not justified in characterizing Toland, if only playfully, as a freethinking magus?

Through his use of Bruno, Toland managed to devise a philosophy of nature that he could effectively posit against the church's Newtonianism. The public confrontation between Toland and Samuel Clarke, and Newton's private philosophizing on the very questions raised by Toland, illustrate the social and political content of the Newtonian ideology more graphically than any of the other confrontations between the church and the freethinkers. Ostensibly Toland and Clarke quarreled about the relationship between matter and motion. In reality they were political enemies who quarreled about the world natural because both possessed profoundly different conceptions of how the world political should operate. In each case their conceptions about the operations of nature and the polity were intimately entwined.

Toland embarked on the path that led to his confrontation with Newtonianism when in 1698 he purchased a copy of Bruno's *Spaccio de la bestia trionfante* from the sale of the library of Francis Bernard. The book was bound together with three other important works by Bruno, *De la causa, principio*

[89] Venturi, 49–55, 64–67.

et Uno, De l'infinito universo et mondi, and *Dialogi la Cena de la Ceneri,* [90] which in their totality contain the essence of Bruno's philosophy and cosmology and also reveal his profound involvement in the Hermetic tradition.[91] At this point it should be recalled that Toland had received a firm foundation in natural philosophy, of a Newtonian sort, from David Gregory and could easily have extracted what he wanted from Bruno's Hermetic philosophy of nature without necessarily accepting the mysticism and magic inherent in it.

Toland reacted enthusiastically to the *Spaccio.* He circulated it among his friends, members of a secret club, and described Bruno as one who "treats all miracles as fables; but he maintains at the same time that the pagan mythology wasn't much more unintelligible or absurd or monstrous than Judaic or Christian theology. He desires that men who are removed from all prejudice in favor of one or the other, or of whatever religion there is, would wish to accept as the rule of their conduct only the law of natural religion."[92] The author of *Christianity Not Mysterious* thought he spied in Bruno another freethinker and was delighted.

But Toland did not confine his use of Bruno to private and secret discussions with his friends. On his trip to the Continent in 1702 he brought the *Spaccio* with him, and in public de-

[90] G. Aquilecchia, "Nota su John Toland traduttore di Giordano Bruno," *English Miscellany,* 9 (1958), 77–86. Frances Yates first called this article to my attention. This discussion of Toland's relationship to Bruno first appears in my "An Interpretative Study of the Life and Thought of John Toland" (M.A. thesis, Cornell University, 1966). For some information on the circulation of Bruno's works in the eighteenth century see MSS ADD 12062 and 12063, B.L.

[91] Yates, *Bruno,* 211–229. As any reader familiar with this book will realize, my interpretation of Bruno relies heavily upon it.

[92] Chaufepié, ed., II, 455; the same letter is reprinted in *Nova Bibliotheca Lubecensis,* VII, 1756, 158. Neither editor reveals where he obtained this very important Toland letter. For evidence that Toland could read Italian fluently, see I. Barzilay, "John Toland's Borrowings from Simone Luzzatto," *Jewish Social Studies,* 30 (1969), 75–81.

bate and private discussions Toland espoused Bruno's natural philosophy.

Toland's interests on the Continent were both political and intellectual. He had accompanied the earl of Macclesfield on the 1701 mission that brought to Hanover the Act of Settlement. Toland's appearance with the delegation is somewhat surprising. His reputation as a heretic was widespread; but apparently this fact did not disturb the faction he represented. Although Toland's tract, *Anglia Libera*, in support of the Protestant succession, is generally cited as the reason for his appointment,[93] it is more probable that he represented the radical faction of Whigs to which he belonged.

On this first visit Toland incurred the suspicion and dislike of Leibniz, who was in Hanover to assist the Electress Sophia and to ensure the success of the English mission. Leibniz suspected that Toland had been sent by "several gentlemen" to learn the workings of the Hanoverian state and that he had republican leanings.[94] In the eyes of the shrewd Leibniz, Toland had "rendered himself odious" during his visit.[95] Yet despite this condemnation, Toland again journeyed in 1702 to the Continent, this time to the court of Prussia, where he obtained an audience with Queen Sophie Charlotte. Upon his arrival Toland met Leibniz again, probably to his annoyance, and offered him greetings from Pierre Bayle, whom Toland had visited during a stopover in the Netherlands.[96]

He spent his time at Berlin in philosophical debate with

[93] N. Luttrell, *A Brief Historical Relation of State Affairs from September, 1678 to April, 1714*, V (Oxford 1857), 67, 100–101, 174; and *DNB*.

[94] C. Gerhardt, *Die Philosophischen Schriften von Gottfried W. Leibniz* (Berlin, 1882), III, 282, written to G. Burnet; also Onno Klopp, *Correspondenz von Leibniz mit der Prinzessin Sophie, 1680–1714* (Hanover, 1873), II, 271–274, passim. Translations are my own.

[95] Klopp, III, 353, Leibniz to Spanheim, Dutch ambassador, June 1702.

[96] Gerhardt, III, 63–64.

Leibniz and the Lutheran theologian, Isaac Beausobre.[97] Toland presented a discourse on the nature of the soul, in which he claimed that the soul was essentially corporeal. Leibniz's impression of Toland's doctrines was that "they revolved almost on the doctrine of Lucretius, that is to say, on the coming together of 'corpuscles,' " but that Toland did not explain how in his system "arose the notion that matter has movement and order; nor does he comment how there is sentience in the world."[89] Leibniz concluded that Toland's system had little value, adding that "instead of keeping himself busy with philosophy, which isn't his *forte*, he would do better to apply himself to the researching of facts."[99] Leibniz dismissed Toland's thought as being little better than fiction and Toland as not capable of producing a serious tract in philosophy.

But a philosophical tract was precisely what Toland did produce. It is quite possible that this tract was an early draft of *Letters to Serena* (1704). But why did Toland choose to reveal it in the hostile atmosphere of the Prussian court? The answer appears to be that he was seeking to convert the queen, who was noted for her interest in religious debates and for her eagerness to learn from them. To accomplish this aim he brought another tract, his own valuable copy of Bruno's *Spaccio* and presented it to the queen for her perusal.[100] Toland apparently avoided displaying the book to Leibniz; there is no mention of it in the letters that describe Toland's visit. In 1709, Leibniz responded to a letter from Toland and enquired whether the title of Bruno's book was not "Specchio" rather

[97] See *Bibliothèque Germanique*, V (VI) (Amsterdam, 1723), 40–47, passim. F. H. Heinemann has seen Toland's contact with Leibniz as proof that Toland's philosophy was influenced by Leibniz. In my opinion, neither the philosophy nor Toland's comments on it supports this interpretation. See F. H. Heinemann, "Toland and Leibniz," *Philosophical Review*, 54 (1945), 437–457, passim.

[98] Klopp, II, 362.

[99] Ibid., 363. The tone of Leibniz's dismissal of Toland resembles that later rendered against Leibniz by Newton.

[100] Erman, 201. Cf. *Bibliothèque Germanique*, V (VI), 50.

than *Spaccio*; he said that he was not familiar with this notorious work by Bruno.[101] Toland responded by promising to explain to him that *Spaccio* and "the Pantheistic notion of those, who believe no other eternal being but the Universe."[102]

At their court meeting in 1702, Leibniz had refuted Toland's philosophy. In a letter to a friend, Leibniz described it as "precisely that of Hobbes, i.e. that there is no other thing in nature but its shapes and movements. This would also be the opinion of Epicurus and Lucretius except for the fact that Epicurus and Lucretius would admit the void, and the atoms or hard particles. Instead Hobbes counts everything to be full and soft, and that is also my opinion," but, he continued, "I believe that above matter, above that which is purely passive and indifferent to movement, it is necessary to seek the origin of action, perception and order."[103]

Leibniz's association of Toland's philosophy with that of Hobbes indicates that Leibniz failed to recognize the similarity between Toland's natural philosophy and that of Bruno. Perhaps Leibniz was unfamiliar with Bruno's Italian works. Furthermore, Leibniz's identification of Toland with Hobbes, despite their patent political antipathy, indicates that Leibniz may have realized the ideological use made of Hobbes within English political and intellectual debates.[104]

Yet Bruno's thought remains the main source for the development of Toland's philosophy. In discussing the relationship between the philosophies of Bruno and Toland it is necessary to rely on only those works by Bruno that Toland is known to have possessed. Among them was the *Spaccio*, in

---

[101] Desmaizeaux, ed., *Toland*, II, 387. Cf. MSS ADD 4465, ff. 5–7, B.L. See Giancarlo Carabelli, "John Toland e G. W. Leibniz: Otto lettere," *Rivista critica de storia della filosofia*, 29 (1974), 412–431. Carabelli has recently finished a complete bibliography on Toland.

[102] Desmaizeaux, ed., *Toland*, II, 394.

[103] Gerhardt, VI, 519.

[104] Q. Skinner, "The Ideological Context of Hobbes' Political Thought," *Historical Journal*, 9 (1966), 286–317, esp. 307.

which we learn that matter, the world, and the infinite universe are in constant motion. Matter exists in a universe that is infinite and eternal and that contains an infinite number of worlds.[105] God gives to matter and the universe an inherent harmony that is, like God, all-pervasive. God is the spirit who is the source of all energy and change in matter; he is entire, infinite, true, all, good, and one.[106] Because matter is infused with God and is in constant motion, Bruno claims that death is merely an illusion. The appearance of destruction is the result of the ever-changing nature of matter. The souls of men "transmigrate from body to body, through various changes and modifications, and . . . they go to dwell in the bodies of pigs, which are the idlest animal in the world; perhaps, indeed, they might become oysters in the sea, which are attached to rocks."[107]

Although Bruno accepted the atomic structure of matter, he did not allow for the void in nature.[108] The medium into which Bruno places the atoms might better be called an aether. The concept of absolutely empty space in the universe, within which Newton allowed for the operation of active principles or spiritual forces, does not appear in Bruno's Hermetic natural philosophy, nor again, a century after him, in Toland's.

Bruno's philosophy of nature must be seen as simply an aspect of his borrowings from the Hermetic wisdom. For him Hermeticism constituted, in effect, a new religion devoid of

[105] Arthur Imerti, trans., *The Expulsion of the Triumphant Beast* (New Brunswick, N.J., 1964), 134.

[106] Ibid., 135–137.

[107] Ibid., 126–127.

[108] Ibid., 137. Paul Henri Michel, *La Cosmologie de Giordano Bruno* (Paris, 1962), 147–148. The opposite and less convincing argument is made by Dorothea Waley Singer, "The Cosmology of Giordano Bruno," *Isis*, 33 (1941), 189. I realize that Newton later introduced the concept of an aether into his natural philosophy. He allowed for an aether that filled up the void between material atoms. I owe this reminder to I. B. Cohen. The Newton-Toland dialogue would not have been affected by this later introduction.

superstition and of reliance on supernatural explanations of either the physical or the human order. He proposed the Hermetic religion complete with its pantheistic materialism as an alternative to both the Reformation and the Counter Reformation. Because of his efforts he found himself condemned as a heretic and was burned at the stake in 1600.

Bruno's influence throughout the seventeenth century, especially in England, remains a complex problem and one that cannot be solved here. What can be said, with certainty, about Bruno in England is that the imprint of his ideas about nature and religion is evident throughout Toland's *Letters to Serena*.

Toland meant his tract as a philosophical essay; but it was also to be a general work with a popular literary form. The form was standard for the day—five conversations on topics of interest written to a lady of leisure and learning. Fontenelle's *Plurality of Worlds* had been cast in this manner and was one of the most famous books of popularized science in the late seventeenth century. Toland modeled *Letters to Serena* on the literary form of the *Plurality of Worlds*.[109] He used its form and style in the hope that perhaps his letters would attain equal popularity. He informs his readers that he intends to show the ability of certain women, like Queen Sophie, to understand the intricacies of philosophy. The *Letters* is not intended to offend, and he desires that the contents of his book should please all.

This preface hardly foretold the contents. The ideas expressed in *Letters to Serena* doomed any prospect of its popularity. The imprint of Bruno's philosophy is evident throughout the book, although the fifth letter reveals most completely the influence of Bruno and the Hermetic tradition.

In the fourth letter, Toland begins his exposition of his own "easy, first, and plain" natural religion with a critique of

[109] See Gerald Meyer, "Fontenelle and Late 17th Century Science," *History of Ideas Newsletter*, 4 (1958), 26. The most popular translation was that done by John Glanvill (1688, 1702).

Spinoza. Toland denounces the detractions which are leveled
against Spinoza, claiming that they are contrary to religion
and common civility. His respect for Spinoza as a defender of
religious toleration does not lead him to modify his opposition
to Spinoza's system.

In view of the general misunderstanding accorded Spinoza's
thought at this period Toland's analysis is perceptive.[110] He
objects that Spinoza has not explained God's relationship to
the universe, nor has he explained the relationship between
matter and motion. Spinoza could not, therefore, "possibly
show how the Diversity of particular Bodys is reconcilable to
the Unity of Substance, or to the Sameness of Matter in the
whole Universe."[111] Because Spinoza has united God with the
universe and not asserted a principle of absolute motion, he
condemns the universe to passivity. Toland asserts that the
universe is in constant motion and that this innate activity
exists and is perpetuated by the inherent motion in matter.
Having established this principle, Toland dismisses Spinoza's
system and turns to a presentation of his own philosophical
system.

In the fifth letter the Brunian philosophy is evident throughout, although missing from this exposition is any clear statement that God and nature are one. But pantheism is evident
throughout Toland's other writings. In this letter Toland
deals with certain philosophical entities—matter, motion, space,
and the void—the definition of which determined the structure
of any early eighteenth-century natural philosophy.

Matter, the world, and the universe are in constant motion:
this motion is inherent in matter and must be included in any
definition of matter. All of the things in the universe are linked
in a great chain and "all these depend in a Link on one another,
so their Matter (to speak in the usual Language) is mutually

---

[110] For a more lengthy discussion of Spinoza's influence on Toland
see Rosalie L. Colie, "Spinoza and the Early English Deists," *Journal
of the History of Ideas*, 20 (1959), 43–46.
[111] *Letters to Serena* (London, 1704), 147.

resolv'd into each other: for Earth, and Water, and Air, and
Fire, are not only closely blended and united, but likewise
interchangeably transform'd in a perpetual Revolution; Earth
becoming Water, Water Air, Air Aether, and so back again
in Mixtures without End or Number."[112] All beings are somehow united and have existed eternally; their union is not static
but dynamic, and all things become one.

This infinity of all things exists in the one, matter, which is
indestructible and in constant motion. The effects of this inherent activity are all the sensible qualities observable in bodies.
Yet amid these multitudinous changes in nature there is a unity
and indestructibility in matter. Nothing is ever finally destroyed; all things eternally merge one into another. "All the
Parts of the Universe are in this constant Motion of destroying
and begetting, of begetting and destroying: and the greater
Systems are acknowledg'd to have their ceaseless Movements
as well as the smallest Particles, the very central Globes of the
Vortexes turning about their own Axis; and every Particle in
the Vortex gravitating towards the Center."[113]

Matter is one. Within this essential unity exists multiplicity
and diversity. All parts of the universe, the smallest particles
and most immense bodies, are in a constant state of motion,
made possible by an inherent force within matter. All space is
filled with matter that moves constantly, impelled by its own
dynamism. Outside, immaterial, forces are unnecessary to
move matter, and therefore the empty space or void within
which matter would be moved is equally unnecessary. As a
result of placing motion inherent in matter, Toland eliminates
the concept of a void. He claims that a void is necessary only
if we conceive matter, as do the Newtonians, as "sluggish" or
inactive, in need of a principle of motion to act on it within a
void. In the philosophies of Newton and of Toland matter
serves very different functions. To Newton and his followers,

[112] Ibid., 187–188.
[113] Ibid., 188.

matter is "brute and stupid," moved by some action at a distance which originates ultimately from God. To Toland matter is the source of life, change, and even order. It is a completely protean force in nature.

With this view of matter, Toland could still maintain that the planets move in their fixed orbits, revolving around their fixed axes, and that the universe is in constant motion. This infinite dynamic universe, operating by its inherent force, exists in relative space and time within which there are "numberless Worlds."[114] By retaining the force of gravity and making it inherent within all bodies, Toland is able to give proof to his notion of the harmony within nature and the universe.

Man has an important place in this cosmos. He, too, is part of nature and intimately bound to his fellow creatures—"nothing is more certain than that every material Thing is all Things and all Things are but one."[115] As such, he is subject to the forces operating within nature. Man's body and soul undergo metempsychosis; when we die we only "cease to be what we formerly were, so to be born is to begin to be something which we were not before."[116]

God's place in this system does not seem to present a problem to Toland. While in the *Pantheisticon* God is made one with nature, in the fifth letter to Serena, Toland includes God at the end of the tract—almost as an afterthought. God is immaterial, and his place in the universe is to give it order.[117]

In arguing his system, Toland addresses Newton directly and attempts to show that Newton's physical laws, as presented in the *Principia*, could be used to support his own unique materialist philosophy. Although he praises Newton as the scientist who has seen most clearly into the actual state of

[114] Ibid., 217.
[115] Ibid., 192.
[116] Ibid., 191.
[117] Ibid., 236.

matter,[118] Toland opposes his own philosophy to the Newtonian natural philosophy. Toland quotes Newton's argument for absolute space and time and then argues that space and time are relative. He claims that Newton's laws need not be interpreted as Newton has interpreted them; they are "capable of receiving an Interpretation favourable to my Opinion."[119]

Throughout *Letters to Serena*, Toland bases his arguments against Newton on an interpretation of gravity which is consistent with his own cosmology. To him gravity is not simply a law which governs the movement of bodies; it also expresses the inherent motion of all bodies.[120] All bodies gravitate to the center of the earth, and this is an expression of their inherent, harmonious activity. Toland objects to the mechanical philosophers because they conceive matter as inactive, moved by the laws of action which are external to the nature of bodies.

Newton discovered the particular forces and figures in nature, but, boasts Toland, "as for the general or moving Force of all Matter, I wou'd flatter myself, that I have done something towards it in this Letter."[121] The Newtonian philosophers claimed, however, that Toland had done exactly what Newton had cautioned against; he framed hypotheses and offered an explanation of nature which was not based on experiment. By using Newton's system to his own advantage, Toland has demonstrated that, if gravity be regarded as a force inherent in bodies, matter is "alive," and that this fact would explain the dynamic and eternal nature of the universe.

The use of Newtonian science to support a hylozoic philosophy posed a threat that could not be allowed to remain unanswered. In the 1704–1705 Boyle lectures, Samuel Clarke publicly led the attack by denouncing Hobbes, Spinoza, and Toland, despite the extreme differences in their philosophies,

[118] Ibid., 201–202.
[119] Ibid., 183.
[120] Ibid., 188.
[121] Ibid., 234.

as atheists. He used in support of his theology the natural philosophical arguments which were subsequently published in the twenty-third query to Newton's Latin *Optice* (1706), translated by Clarke and later to become the major part of the thirty-first query (1717–1718). In this query Newton asserts that the only method suitable for an explanation of nature is attained by reasoning from experiments and observations and thus, by induction, arriving at general conclusions. The understanding of particular effects leads to a comprehension of their causes. Then, by a method of composition, these causes are tested by experiment in order to arrive at a more general cause or rule, from which particular effects might in turn be demonstrated. Simply stated, "Hypotheses are not to be regarded in experimental philosophy." When Toland had philosophized about the nature of matter and the cosmos not from experiment but from a predetermined set of principles, he like the other atheists had violated the rules of right reasoning.

Likewise Toland's assertion that motion is inherent in matter denied one of the most basic principles of the Newtonian natural philosophy. In response to philosophies that were materialistic or that encouraged materialism Newton explained that bodies are at rest until moved or that they remain in motion depending on the degree of force initially applied: "The *vis inertiae* is a passive Principle by which Bodies persist in their Motion or Rest, receive Motion in proportion to the Force impressing it, and resist as much as they are resisted. By this principle alone there never could have been any Motion in the World."[122]

Motion is the result of the active principles present in the universe, and in the Newtonian system these have their source in the omnipotence of God, for whom space exists as a sensor-

[122] Newton, *Opticks: or a Treatise of the Reflections, Refractions, Inflections and Colours of Light* (2d ed., with additions, London, 1718), 372–373.

ium.[123] By this phrase Newton took up what had earlier been a concept in the natural philosophy of Henry More, the Cambridge Platonist. For Newton space is virtually an attribute of God, and through this intimate relation with the universe he is the source of its activity and order.[124] God must constantly act in this universe, for the motion that "we find in the World is always decreasing, there is a necessity of conserving and recruiting it by active Principles, such as are the cause of Gravity . . . . For we meet with very little Motion in the World besides what is owing to these active Principles."[125] Thus God is essential for the maintenance and preservation of the Newtonian universe. He is the "ever-living Agent, who being in all Places, is the more able by his Will to move the Bodies within his boundless uniform Sensorium."[126] If motion be inherent in matter, God is rendered impotent and therefore nonexistent. To Clarke and Newton, Toland's system was not simply wrongheaded, it was atheistical.

In the Boyle lectures Clarke bases his refutation of atheism on the moral assertion that it denies free will. Those philosophies that deny it is "possible to Infinite Power, to Create an Immaterial Cogitative Substance, indued with *a Power of beginning Motion*, and with a *Liberty of Will or Choice*" are "of the greatest Consequence to Religion and Morality."[127] In the Newtonian natural philosophy, Clarke found the principles essential to theism and to a guarantee of man's free will;

[123] Ibid., 375, 379.

[124] Henry More, *A Collection of Several Philosophical Writings* . . . (London, 1712), 199–201. Cf. A. Koyré and I. B. Cohen, "The Case of the Missing *Tanquam*: Leibniz, Newton and Clarke," *Isis*, 52 (1961), 561–563.

[125] Newton, 375.

[126] Ibid., 379.

[127] S. Clarke, *A Demonstration of the Being and Attributes of God* (London, 1705), 159–160. Cf. John Gay, "Matter and Freedom in the Thought of Samuel Clarke," *Journal of the History of Ideas*, 24 (1963), 85–105, passim.

he found an alternative to the philosophies of Hobbes and Spinoza, but more immediately and urgently to the natural philosophy proposed by Toland.

Clarke defines God as a being who is incomprehensible, self-sufficient, eternally existent, infinite, omnipresent, immense, and full. This definition precludes the eternity of matter and makes it dependent on the will of God. The motion residing in matter is the result of God's omnipotence as he makes operative the laws of physical motion. The "admirable order" present in the universe is the manifestation of the will of an intelligent and all-powerful being.[128] In order for matter to satisfy the laws of motion, a void is essential to the universe. Matter is extended and impenetrable; although it could be boundless, its motion is not inherent. God moves matter either directly, by his will operative as the law of motion, or by creating an "Immaterial Cogitative Substance [man], indued with a power of beginning motion, and with a Liberty of Will or Choice."[129] Specifically in answer to Toland, whom he cites in the margin of his text, Clarke argues by *reductio ad absurdum* that if motion be inherent in matter the "*Conatus* equally to move every way at once, is either an absolute Contradiction, or at least could produce nothing in Matter but an Eternal Rest of all and every one of its Parts."[130]

When, in 1704–1705, Clarke preached against Toland, he used Newton's ideas on the relationship between matter and motion, the existence of the void, and God's role in the universe. The twenty-third query to the Latin edition of the *Optice* offered ample support to Clarke's theology. And the recently discussed private papers of Newton[131] also confirm the similarity between Clarke's and Newton's metaphysical

---

[128] Clarke, 118–119, 125.

[129] Ibid., 159.

[130] Ibid., 47.

[131] J. E. McGuire and P. Rattansi, "Newton and the Pipes of Pan," *Notes and Records of the Royal Society*, 21 (1966), 118.

views. Sometime between the publication of the English edition of the *Opticks* in 1704 and the Latin edition in 1706, Newton composed a draft version of the twenty-third query that differed from the published text. This other version clearly indicates Newton's concern with the relationship between motion and matter and his opposition to current philosophies that claimed an inherent motion in matter. It seems reasonable to conclude in light of Toland's use of Newton and of Clarke's specific refutation that Toland's philosophy, rather than any other version of materialism, must have been uppermost in Newton's mind when he wrote this draft of the twenty-third query.

Newton had been mentioned by name in the *Letters to Serena*, and Toland had used the law of gravitation to support his own materialistic philosophy. Although ever anxious to avoid theological debate, Newton must have been alarmed at the attack made on his physics as well as by the use to which it was put, by Toland, a former student of David Gregory and an associate of radical political and philosophical circles in England and on the Continent. Furthermore, Newton truly believed that his science supported Clarke's refutation of atheism and materialism.

This draft version of the twenty-third query begins: "By what means do bodies act on one another at a distance?" Newton's unpublished answer reveals his belief in a *prisca theologia* and his rather startling views on the relationship between God and the universe. It also demonstrates Newton's deep concern for the theological and religious implications of his science, a concern that is also expressed in the earlier "classical" scholia. Newton's writing of these draft scholia in the 1690s and the similarity of the ideas expressed in the following query with those contained in the scholia indicate that his interest in the ancient origins of his philosophical wisdom, as well as in its implications for religion, was a long-standing and important one.[132] Because of the relevance of this document to the

[132] Ibid., 108–121.

philosophical and religious debates of Newton's time, portions of it must be quoted at length:

By what means do bodies act on one another at a distance. The ancient Philosophers who held Atoms and Vacuum attributed gravity to atoms without telling us the means unless perhaps in figures: as by calling God Harmony & representing him & matter by the God Pan and his Pipe, or by calling the Sun the Prison of Jupiter because he keeps the Planets in their Orbs. Whence it seems to have been an ancient opinion that matter depends upon a Deity for its laws of motion as well as for its existence. . . . These are passive laws and to affirm that there are no other is to speak against experience. For we find in ourselves a power of moving our bodies by our thought. *Life and will are active principles by which we move our bodies, and thence arise other laws of motion unknown to us.*

And since all matter duly formed is attended with signes of life and all things are framed with perfect art and wisdom and nature does nothing in vain; if there be an universal life and all space be the sensorium of a thinking being who by immediate presence perceives all things in it, as that which thinks in us, perceives their pictures in the brain: these *laws of motion arising from life or will may be of universal extent.* To some such laws the ancient Philosophers seem to have alluded when they called the God Harmony and signified his actuating matter harmonically by the God Pan's playing upon a Pipe and attributing musick to the spheres made the distances and motions of the heavenly bodies to be harmonical and represented the Planets by the seven strings of Apollo's Harp.[133]

The above philosophical argument is essentially the same as Samuel Clarke's. God is the author of the laws of motion. His

[133] MSS ADD 3970, fols. 619ʳ, U.L.C. This manuscript was first made available to me in 1965–1966 through the kindness of Henry Guerlac. Italics are my own. For a different approach to this manuscript see J. E. McGuire, "Force, Active Principles, and Newton's Invisible Realm," *Ambix*, 15 (1968), 154–208. See also my "Interpretative Study."

will operates the motion in the universe. But in answer to Toland's charge that the Newtonian universe is "dead" or "sluggish," Newton reveals his belief that all matter "is attended with signes of life." Because all space is the sensorium of God, the active principles of life and will are omnipresent, and thus there is a universal life in nature. Perhaps the reason why Newton chose not to publish this startling version of the twenty-third query was his realization that his views on the relationship between God and the universe could easily be attacked for their unorthodox tendency or be used by those who held to a mystical philosophy of nature. We see that his fears were well grounded when we recall Leibniz's later attack on this very notion of the "sensorium of God," as found in the twenty-third query to the Latin *Optice*, and his assertion to Clarke that God must be *supermundanus*. In his dialogue with Clarke, Leibniz rightly assumed that Newton had received his notion of God's spatial attribute from Henry More.[134]

Earlier, in a letter to Toland (1709), Leibniz had asserted that God is *intelligentia supermundana* and used this argument against Toland's notion that God and the universe are one.[135] It is possible that in the famous correspondence of 1715 when Leibniz objected to Clarke about Newton's conception of God's spatial relation, he feared a connection between Toland's philosophy and Newton's easily misinterpreted notions about the relationship between God and nature.

In his first letter to Clarke, Leibniz began by expressing his concern that "natural religion itself, seems to decay (in England) very much. Many will have Human *Souls* to be material: Others make *God himself* a corporeal Being."[136] Leibniz pro-

---

[134] Cf. A. Koyré and I. B. Cohen, "Newton and the Leibniz-Clarke Correspondence," *Archives internationales d'histoire des sciences*, 15 (1962), 86–94.

[135] MSS ADD 4465, ff. 5–7, B.L.; cf. Desmaizeaux, ed., *Toland*, II, 383–387.

[136] S. Clarke, *A Collection of Papers which Passed between the Learned Mr. Leibniz and Dr. Clarke* . . . (London, 1717), 3.

ceeded to list his objections to the Newtonian natural philosophy, which in his opinion was a factor in the decaying state of English religion. Clarke's pointed reply laid blame for this state on the materialists (among whom he would have included Toland) and asserted that Newton's "Mathematical Principles of Philosophy are most directly repugnant" to their false philosophy.[137]

Because of his distance from events in England, Leibniz may have been led, possibly by Toland's boastful claims, to fear a similarity between the ideas of Newton and the pantheistic philosophy of Toland and his friends. But to English churchmen no such association would have been possible. The philosophies of the freethinkers bore no resemblance to the serious speculations of Newton. They were seen as radical propagandizers whose efforts threatened the church and all revealed religion.[138]

Toland's activities became most dangerous to the church and its political interests during his trip to Hanover, then in the publication of *Letters to Serena*, and later in his involvement in the translation and publication of Bruno's *Spaccio* (1713). Although it has been claimed that Toland was the translator of the *Spaccio*, evidence would seem to show that William Morehead actually translated it for the private use of Anthony Collins.[139] Following the sale of an Italian copy of the *Spaccio* at a London auction in 1711 and the public attention given to the high price obtained for the book,[140] Toland took the translation from Collins' library and sent it to the press. This English copy also brought an excellent price. In his diary, Thomas Hearne recorded that the auctioned copy of the *Spaccio* had been from the library of Charles Bernard,

[137] Ibid., 9.
[138] Cf. P. Casini, "Toland e l'attività della materia," *Rivista critica di storia della filosofia*, 22 (1967), 24–53.
[139] Dorothea Waley Singer, *Bruno, His Life and Thought* (New York, 1950), 192n.
[140] *The Spectator*, V, no. 389 (London, 1712), 301–305.

and it "was sold for twenty-seven Pounds, being bought by one Mr. Clavell of the Middle-Temple, a great crony (unless I am misinform'd) of Toland, Stevens, Tyndale and other Atheisticall and ill Men."[141]

Two years later Hearne recorded the existence of a group of freethinkers, among them Collins and the then repentant Peter Needham (1680–1731), all of whom could be identified by their speaking favorably of Bruno's *Spaccio*.[142] Collins made references to this group in *A Discourse of Freethinking*, and his own views on natural philosophy found in that discourse bear a striking resemblance to those found in the *Spaccio*.[143] He was sternly answered by Bishop Ibbot in the Boyle lectures for 1713–1714. In these influential sermons Ibbot, following Clarke's pattern, attacked the freethinkers and included in his remarks a discussion of Bruno's *Spaccio* and the attitude to religion it contained.[144]

At the beginning of the eighteenth century, a coterie of English freethinkers attacked all revealed religion, used the writings of Bruno supplied by Toland, and probably espoused the natural philosophy found in *Letters to Serena*. Although it is difficult to determine the exact relationship of this group with Continental thinkers of the period, Toland's unpublished notes reveal that he circulated at least three clandestine manuscripts on the Continent. They were entitled "Life of Giordano Bruno," "Translation of Bruno's Asse, 2 Dialogues," and "Bruno's Sermon."[145] One of these may have become a source

[141] T. Hearne, *Remarks and Collections* (Oxford, 1906), III, 202, Aug. 7, 1711.

[142] Ibid., IV, 172, May 6, 1713.

[143] Collins, *A Discourse of Freethinking* (London, 1713), 47–48, 119, 150–151. Cf. J. H. Broome, "Une Collaboration: Collins et Desmaizeaux," *Revue de littérature comparée*, 30 (1956), 160–179, passim.

[144] Benjamin Ibbot, *A Course of Sermons Preach'd for a Lecture Founded by the Hon. Robert Boyle, Esq. . . . Wherein the True Notion of the Exercise of Private Judgement, or Freethinking, in Matters of Religion is Stated* (London, 1727), 40–54.

[145] MSS ADD 4295, 43, B.L. This list is printed in F. H. Heinemann,

for a work later published by members of Baron d'Holbach's salon and entitled *J. Brunus redivivus ou traité des erreurs populaires* (1771). Toland's collected works, published by Pierre Desmaizeaux, also contain an English translation of the introductory epistle to Bruno's *Of the Infinite Universe and Innumerable Worlds.*

In 1720, Toland finally revealed in the *Pantheisticon* the philosophy and ritual of the society to which he claimed membership. This difficult-to-obtain and anonymously published book displays the complete dominance of the hylozoic natural philosophy derived from Bruno. The pantheists believed that "all things are from the whole, and the Whole is from all things . . . they assert that the Universe . . . is infinite both in Extension and Virtue, but one, in the Continuation of the Whole, and Continuity of the Parts. . . . Finally, whose integrant Parts are always the same, and constituent Parts always in Motion."[146] The doctrines that the universe is infinite and contains innumerable worlds, that all things are united into one, and that matter and the universe are in constant motion, as well as in constant repose, had been first revealed by Toland in *Letters to Serena.* The important element of Bruno's thought not presented, however, in *Letters to Serena* is the complete pantheism of the *Pantheisticon;* "eminent reason" is now defined as "God, whom you may call the *Mind* if you please, the *Soul* of the Universe."[147] In Toland's system God gives force and energy to the universe and by his existence as the informing spirit of the universe gives harmony to its various parts. It is this belief that allows Toland to call the members of his Socratic Brotherhood, pantheists.[148] Their pantheism is in agreement with the philosophy of the ancient

---

"Prolegomena to a Toland Bibliography," *Notes and Queries,* 185 (1943), 184. Cf. Ira O. Wade, *The Clandestine Organization and Diffusion of Philosophic Ideas in France from 1700 to 1750* (New York, 1967, 1st pub. 1938), nos. 48, 58, 80.

[146] Toland, *Pantheisticon,* 15.

[147] Ibid., 17.

[148] Ibid., 17–18.

"Pythagorics," who according to Hermetic lore were Hermetists.[149] The religion of the pantheists is esoteric; it is meant to be practiced in secret by the members. Perhaps with the hope of acquiring followers, Toland reveals in the *Pantheisticon* a part of their philosophical liturgy. It provides ample illustration that by 1720, Toland and his associates had abandoned public debate; their activities had devolved into private rituals and philosophical aphorisms.

In view of other supporting evidence found in Toland's manuscripts little reason seems to exist for doubting his claim to membership in the esoteric society that used the secret ritual and espoused the natural philosophy described in the *Pantheisticon*. This group was undoubtedly quite distinct from the Grand Lodge of 1717, of which Desaguliers, a faithful Newtonian, was grand master and which included a sizable number of fellows of the Royal Society.[150] The attacks launched by the church against these freethinkers, with the Newtonians at the vanguard of the assault, indicate that, if only for a brief time, Toland and his Socratic Brotherhood appeared to offer a viable and dangerous alternative to the politics and natural religion practiced and preached by the post-Revolution church.

The antagonism between the pantheistic materialism of Toland and his friends and the natural philosophy of Newton and his coterie reveals as graphically as possible the political and ideological uses to which natural philosophy could be put in the late seventeenth and early eighteenth centuries. Presumably Toland found in pantheism justification for a utopian republic, free from clerical influence and religious intolerance, where intellectual and political freedom amounted to a loosely defined social equality. All men were equally a part of nature,

[149] Yates, *Bruno*, 131, 185.

[150] Wright et al., ed., *Gould's History of Freemasonry throughout the World* (New York, 1936), I, 262; cf. Jean Barles, "Le Schisme maçonnique anglais de 1717," *Les Archives de Trans-en-Provence*, 61 (1937), 286–287, for a different interpretation.

and churches or monarchs counted for far less when spiritual transcendence amounted to the merger of the one into the all. Conversely, the Newtonians argued for social control, the preservation of Church and king, and the pursuit of self-interest by acquisitive men who vie for control over matter and nature. The public interest and civic virtue survive as practical ideals only when men acknowledge the harmony of the mechanical universe controlled by the same providence who directs the natural leaders of society and government.

In this political and philosophical clash with the freethinkers we can pronounce the Newtonians victors. England never became a republic of pantheists, and by 1720, Toland and his friends had retreated into their private world. In the letter to Barnham Goode discussed earlier Toland explained his ritual and liturgy; he also remarked concerning his lack of involvement in public affairs, "I enjoy as profound a tranquility, as if living in Arabia."[151] Despite this retreat, the church never relaxed its vigil against the freethinkers, perhaps because it knew that they symbolized most dramatically its declining political authority. In an effort to reverse that process churchmen turned to natural religion, a rational Christianity, supported whenever possible by science, which could embrace the widest possible audience.

But, as the freethinkers demonstrated, the philosophical props of that natural religion could be strenuously challenged. As churchmen unrelentingly produced new arguments, or repeated old ones, in support of natural religion, they completely obscured those utopian, millenarian sentiments which had been at one time such a vital part of the latitudinarian sensibility.[152] Natural religion, to put it perhaps too boldly, meant simply a practical creed intended to curb the worst excesses of self-interest and acquisitiveness. The latitudinarians ignored religious sentiment or emotion; indeed, when coupled with desire for social and political reform the latitudinarians

[151] MSS ADD 4295, f. 39, B.L.
[152] See Chapter 3.

regarded religious emotion as dangerous to the established order, and they regarded it as being just as pernicious as free-thinking.

Ironically, and revealingly, freethinkers and churchmen in the early eighteenth century did find common agreement over one issue. Both condemned religious radicals, enthusiasts, as opponents of rationalism and as disruptors of tranquility in the state. Shaftesbury's *Essay Concerning Enthusiasm* (1708) was one of many treatises, most by churchmen, written in condemnation of the French prophets,[153] who, as we shall see, led a popular religious movement that sought to institute a millenarian paradise. For all of their mutual hostility neither churchmen nor freethinkers could countenance, for different reasons, a religious movement spawned by persecution that spoke to the disillusioned and disaffected. Their political and social aspirations had no place in the Newtonian universe, nor had their religion a place in the republic of freethinkers.

[153] Anon., *Free Thoughts upon the Discourse of Free-Thinking* (London, 1713), 6, accuses the freethinkers of offering by their ideas justification for the French prophets.

CHAPTER 7

The Opposition:
Enthusiasts

If millenarianism survived in Anglican circles well into the early eighteenth century, it appears safe to assume that at the same time within popular culture and religion significant traces of radical millenarianism lay barely beneath the surface. We know so little about popular culture in this later period, however, that it would be foolish to speculate about the actual extent or seriousness of such beliefs. The reactions of the the established authorities in both church and state to the reappearance of religious enthusiasm during the reign of Anne tells us that any such millenarian movement still provoked immediate fear and reprisals. Even by the early eighteenth century millenarianism of a radical political character could find supporters among the populace in London and in the provinces. The following accorded to the French prophets, a millenarian and enthusiastic group who arrived in London in 1706, attests to the survival of millenarianism in English religious life, and the writings of some of the prophets affirm that millenarianism could still be politically radical. The purpose of this chapter is to analyze the activities of the prophets primarily in relation to the response they elicited from churchmen.

Popular millenarianism survived in England during the Restoration despite the church's firm and rather effective efforts to destroy it. In 1688–1689 prophetic treatises came from the presses in support of William's invasion and its portentous meaning as a stage in the demise of antichrist.[1] We

---

[1] John Partridge, *Annus Mirabilis, or Strange and Wonderful Pre-*

can only presume that the authors of these tracts knew their audience. Some evidence of millenarian activity during the 1690s does survive; in 1694, Evelyn fretted about the followers of John Mason in Buckinghamshire who, he feared, might move on to London.[2] In the same year one Mary Holms recorded that Mason was a new prophet to whom Christ had appeared, that he was bent upon preparing his people for a second coming, and that those who "have faith in Christ's coming are mean unlearned and contomtable [sic] persons in ye Eyes of ye World."[3] The Muggletonians were still active at the time of their founder's death in 1698, and over 240 persons attended the funeral of Lodowick Muggleton.[4] Such sectarian and Anglican millenaries as Mason and Thomas Beverley did indeed prophesy[5] in the 1690s, and if the authorities allowed them to slip through their net, they would not make the same mistake with the French prophets only a few years later.

In Anglican circles there had always existed considerable concern for the fate of French Protestants.[6] That concern arose from both humanitarian and millenarian motivations. English Protestants, although priding themselves on the longevity of their Reformation, beginning, as they claimed, with

dictions and Observations Gathered out of Mr. J. Partridge's Almanack, 1688 (London, 1689, printed in Holland); _Merlini Liberata Errate: Or the Prophecies and Predictions of J. Partridge . . . 1690_ (London, 1692), intended to discredit him as a republican; Anon., _A Modest Enquiry into the Meaning of the Revelations; in a Letter to all Such as Wait for the Kingdom of Christ_ (London, Michaelmas Term, 1688 [very rare]). A good starting point for an analysis of popular religion in this period is Charles J. Sommerville, "Popular Religious Literature in England, 1660–1711: A Content Analysis" (Ph.D. diss., University of Iowa, 1970).

[2] E. deBeer, ed., _The Diary of John Evelyn_ (Oxford, 1955), V, 178.

[3] MSS ADD 34274, f. 142, B.L.

[4] G. C. Williamson, _Lodowick Muggleton_ (privately printed, 1919).

[5] Christopher Hill, "John Mason and the End of the World," in _Puritanism and Revolution_ (London, 1958), 323–336.

[6] See pp. 120–121.

John Wycliffe, nonetheless acknowledged the historical importance of French medieval heresies and the fate of their upholders. Believing by the late seventeenth century that France, more than any other country, upheld Catholic power in Europe, our churchmen watched eagerly for any sign that Protestantism survived and grew strong within the very bastion of antichrist. In 1696, Lloyd wrote to Tenison about the Vaudois Protestants: "They are indeed now in the hands of their bloody enemies, and nothing can preserve them from destruction but some extraordinary appearance of the hand of God. And what that can be, I cannot guess. But if they are destroyed, it breaks my whole scheme of Interpretation of the Prophecies of Scripture. Let that be as it please God."[7] The fate of French Protestantism figured closely in the millenarian speculations of churchmen, and, coupled with the church's general concern for persecuted Protestants of whatever sort, the church's reaction to the French prophets assumes particular significance. For the latitudinarian churchmen repudiated the prophets and all that they sought, and they did so for reasons that tell us much about their natural religion, supported as it was by Newtonianism.

The prophets assume special significance in any discussion of the social context of Newtonianism if for no other reason than that one of their important devotées was Nicolas Fatio de Duillier, mathematician and for a time intimate confidant of Sir Isaac Newton. In early 1690, Fatio became one of the first of Newton's followers to spread his master's views on the Continent.[8] He meditated and wrote extensively on Newton's theory of universal gravitation,[9] and even a cursory reading of

[7] MS 930, f. 42, Lambeth Palace Library.

[8] Frank Manuel, *A Portrait of Isaac Newton* (Cambridge, Mass., 1968), 195–196.

[9] See Bernard Gagnebin, ed., "De la cause de la pesanteur. Mémoire de Nicholas Fatio de Duillier présenté à la Royal Society le 26 février 1690. Reconstitué et publié avec une introduction," *Notes and Records of the Royal Society of London*, 6 (1949), 105–160; and MS francais 603, Bibliothèque Publique et Universitaire de Genève.

his manuscript remains reveals that for him the Newtonian philosophy embodied a new, yet ancient, wisdom that seemed to unlock the very secrets of the universe. Fatio,[10] like Newton, dabbled in the cabalistic tradition, but unlike Newton, Fatio took his enthusiasm for mysticism to a logical and socially dangerous conclusion.

The partially mutilated correspondence between Fatio and Newton, only recently discovered, reveals that as early as October 1689 and undoubtedly slightly before that date, they were engaged in a warm and open communication.[11] That friendship grew in intensity with Newton at one point offering to give Fatio a financial allowance.[12] Both shared interests in mathematics and prophecy, and in January 1693, Fatio wrote to Newton: "But I am persuaded and as much as satisfyed that the book of Job, allmost all the Psalms and the book of proverbs and the history of the Creation are as many prophecys, relating most of them to our times and to times lately past or to come."[13] Fatio explained that the story of creation had a special meaning for the present if understood by seeing the serpent as the Roman Empire, Adam as the clergy, and Eve as the "church or people submitted to the clergy."

Newton's response to Fatio's interpretation indicates his approval of his friend's insights, but adds an interesting note of caution: "But I fear that you indulge too much in fansy in some things."[14] At the time when Newton responded to Fatio's interest in the prophecies—an interest that, of course, Newton

[10] MS français 603, ff. 34–35, and for his interest in Raymond Lull, MS français 605, Bibliothèque Publique et Universitaire de Genève.

[11] H. W. Turnbull, ed., *The Correspondence of Isaac Newton*, III (Cambridge, 1961), 45.

[12] Ibid., 263, 391; cf. Manuel, chap. 9.

[13] Turnbull, ed., III, 242.

[14] Ibid., 245. The relationship between Newton's philosophy of nature and Fatio's religious sensibility has been intelligently discussed in Charles A. Domson, "Nicolas Fatio de Duillier and the Prophets of London" (Ph.D. diss., Yale University, 1972).

shared—he could hardly have imagined that his friend, with a band of French prophets, would translate this concern for the prophecies into an enthusiastic religious movement.

The teachings of the prophets were one logical conclusion to be drawn from the millenarian vision shared by liberal Anglicans. On the basis of that vision, they attempted to interpret the Scriptural prophecies in relation to the future of European Protestantism. More specifically, the efforts of churchmen like Whiston, Allix, Fowler, and Lloyd were meant to interpret the Continental wars in light of prophecies about the destruction of the beast or antichrist. During the reign of Anne, William Lloyd was a constant visitor at the court, and on those visits he advised the queen about her conduct of the French war. He based his advice on his reading of the Old Testament prophecies.[15] A record of Lloyd's prophetic insights occurs in a memorandum by David Gregory dated March 4, 1703/4:

Dr. Lloyd Bishop of Worcester told me at Oxford that all persons agreeing that the coming of Anti-Christ is from the Destruction of the Roman Empire, which at latest was in the time of Augustulus An. Aerae Christ 476 . . . the destruction of Anti-Christ must, at least be An. Aerae Christian 1736. At which time Rome shall be burnt, the papacy destroyed, and Jerusalem rebuilt. This shall be proceeded with the destruction (at least humiliation) of the French Empire & the Conversion of France from Popery. [A verse of the Revelation hangs over the King of France's head.][16]

Lloyd's prophecies were meant to be extremely exact, and he did not avoid the most precise predictions. In 1707, Tenison wrote to inform the Electress Sophia of Hanover that God was at work in the war on the Rhine.[17]

In another memorandum by Gregory we see that at the

[15] A. Tindal Hart, *William Lloyd 1627–1717* (London, 1952), 166–178.
[16] W. G. Hiscock, *David Gregory, Isaac Newton and Their Circle* (Oxford, 1937), 16. Bracketed material added later.
[17] Stowe MS 223, f. 107, B.L.

same time, Fatio de Duillier and his followers also offered precise calculations based on their reading of the Scripture. The memorandum is dated October 26, 1706:

I was in the company into which Mr. Fatio introduced three [?]. Those are they who pretend to inspiration and prophecy. . . . They read a paper. Mr. Fatio has written most of what they have prophecized, and even their prayers, and seemes much taken up with them. He says that the 3rd day after the Battle of Turin, this Cavalier in a Fitt, said that very shortly they should hear News of a greater defeat than any yet happened to the French Kings forces which he has attested by himself and several others hands at the very time when he said it. He prophecys too that in the year 1707, and also in 1708, the Fr. King shall be beat than he was this year [sic].

The constant tenor of their Prophecys is of the Peace of the Church to follow upon the ruin of Rome. They talk of the Restauration of the Church of France as at hand.

They tell of illiterate people with them having the Gift of Tongues, of being in Fire and not being hurt.[18]

Gregory's memorandum reveals that the concern of Fatio and the prophets was with the destruction of the French king, whom they identified as the beast of the Old Testament, and with the continued reformation of Protestantism by the conversion of Catholic opponents. This concern differs little from the concerns of the latitudinarians who preached the defeat of France as one of the steps necessary in the completion of the Reformation. Yet the prophets were labeled as enthusiasts by churchmen and moderate Dissenters such as Edmund Calamy; and in 1707 they were prosecuted for publishing seditious material and for causing public disturbances.[19] This

---

[18] Gregory MS 247, f. 63, The Royal Society.

[19] See Calamy, *Enthusiastick Imposters, No Divinely Inspir'd Prophets* . . . (London, 1707); Anon., *An Account of the Tryal, Examination and Conviction of Elias Marion, and Other of the French Prophets at the Queen's Bench Bar, on 4th of July, 1707, at Guild-Hall, before the Right Honourable Lord Chief Justice Holt; for Publishing False and Scandalous Pamphlets; and Gathering Tumultous As-*

response to the activities of the prophets should become understandable as we discuss the background to their movement,
their teachings, and the excitement generated in London by
their activities.

After the revocation of the Edict of Nantes in 1685 the
plight of the French Protestants had been a constant concern
to the church in England. In the reign of James II, their persecution and subsequent flight from France were seen as ominous signs of the fate awaiting Protestants at the hands of a
Catholic monarch. Even after the Revolution, Lloyd and Sancroft met to discuss the fate of their French brethren and,
as we recall, to interpret their plight in terms of the prophecies.
The Huguenots left France for Switzerland, Holland, and
England where they established parishes and churches, most
notably in London. Huguenot groups, such as the church of
the Savoy, existed almost as independent communities aided
occasionally by the beneficence of the church of England.
In the propaganda of the church, England became a refuge
for the oppressed French Protestants and, by and large, their
immigration was encouraged by churchmen.

In France, the situation of the Protestant minority became
increasingly desperate, and by the turn of the century, virtual
civil war broke out between Huguenots and Catholics led by
Royal troops. This atmosphere of war and persecution nour-
ished a religious movement of visionary prophets who claimed
they had been inspired by God to lead the Huguenots to victory over their Catholic oppressors. The development of this
kind of religious, military phenomenon is hardly surprising
when we recall the writings of Protestant leaders such as
Pierre Jurieu. In the *Accomplishment of the Scripture Prophecies* (1687) he united his plea for victory over Louis XIV
with a millenarian vision that argued for the inevitability of his
demands. Jurieu's wife later joined with the prophets.[20] His

semblies (London, n.d.). Cf. Georges Ascoli, "L'Affaire des prophètes
français," *Revue du dix-huitième siècle*, III (1916), 85–109.
[20] E. de Budé, *Lettres inédites adressées de 1686 à 1737 à J. A. Tur-*

extremism disturbed other Protestant thinkers, such as Pierre Bayle, because they feared the ends to which this millenarian Protestantism could lead.

In 1701 the young Abraham Mazel (b. 1677), son of a Protestant laborer in the Cévennes, claimed to have had a vision wherein he was instructed to lead his people in a campaign against the forces of the king. At about the same time and in the same district, a vision also came to Elias Marion, who then joined with Mazel to aid the Camisards, as the armed bands of Huguenots were called, in their war against religious persecution.[21] That revolt eventually failed, but from August 1703 to May 1704, devastation was wrought in the Cévennes in response to the uprising.[22] As the rebels gradually capitulated, the prophets with their followers fled, and Marion went to Geneva. After a series of further visions and a return to the Cévennes, Marion left the Continent and arrived in England in September 1706. Mazel did not follow him until 1708,[23] but in London, Marion had no difficulty meeting other French refugees or residents who had had a similar religious experience. Most prominent in the prophetic movement were Durand Fage, Jean Daudé, Fatio de Duillier, Charles Portales, Jean Cavalier of Sauve, and Jean Allut. Marion, Fage, Fatio, and Portales became the apparent leaders and spokesmen of a

---

rettini (Paris, 1887), III, 308. Cf. E. Labrousse, Pierre Bayle (The Hague, 1964), I, 231n.

[21] For a general account of the Camisard movement, see C. Almeras, La révolte des Camisards (Paris, 1959); A. Ducasse, La Guerre des Camisards (Paris, 1962); P. Joutard, ed., Journaux Camisards (1700–1715) (Paris, 1965).

[22] For a much fuller account see C. Bost, ed., "Mémoires inédites d'Abraham Mazel et d'Elias Marion sur la guerre des Cévennes, 1701–1708," Huguenot Society of London Publications, 34 (1931); and E. G. Léonard, Histoire générale du Protestantisme (Paris, 1964), III, 12–22.

[23] Bost, 159. Cf. Philippe Joutard, "Les Camisards: 'Prophètes de la Grande Revolution' ou derniers combatants des guerres de religion?" L'Esprit republicain, Colloque d'Orleans, 4 et 5 septembre, 1970 (Paris, 1972).

movement that at its peak included about sixty known recipients of "the vision."[24] Among Fatio's manuscripts in Geneva is a list of the prophets and followers, which includes many French refugees, some English Quakers and Anabaptists, and even Edward Fowler, bishop of Gloucester, and his family. Given his millenarian views Fowler's inclusion is only mildly surprising. Well over three hundred followers are listed.[25]

The prophets and their followers aimed not only to enlist English support for their French campaign, but also to assert "that God in his Love through Christ Jesus, will restore to Mankind what Man Lost by his Corruption and Degeneracy, viz, That the whole Creation shall appear in its *primitive* Beauty, and Man regain the perfection of *Adam*. And his immediate Communion with God."[26] Upon this restoration of mankind the prophecy of Daniel will come true: the beast shall be destroyed and there shall arise "a new heaven and a new earth." To accomplish their aim, the prophets intended to show that the people had been misinformed by their priests who were followed unwittingly and who had taught the Gospels as metaphor and therefore obscured their full and immediate truth.[27]

The prophets chastised not only the ministers of the Huguenot churches, but also the entire clerical structure of the English church. They never confined their preaching and public meetings to the Huguenot communities; they sought and found English converts.

Although millenarianism had been an ever-present aspect of English religious thought in the seventeenth century, by late century most of the Dissenting sects had lost the enthusiasm that had been an integral part of their earlier fervor. Noncon-

[24] Anon., *An Impartial Account of the Prophets: In a Letter to a Friend* (London, 1708). This figure supplied by Hillel Schwartz, to whom I am grateful for his criticisms of this chapter.
[25] MS français 603, alphabetized list, with tribe designations, Bibliothèque Publique et Universitaire de Genève.
[26] *An Impartial Account*, 19.
[27] Ibid., 4.

formists assumed a growing respectability and prosperity, and
blatant millenarianism became identifiable only among Anabaptists, Philadelphians, and certain Quaker sects.

Such sects were a constant source of annoyance to moderate Dissenters who feared the accusation of guilt by association and to churchmen who feared the divisive power of
enthusiastical sectarianism. In 1697, Simon Patrick wrote to
the rector of Dodington, a Rev. Mr. Williams, the following
encouragement: "You have done very worthily and prudently
in stopping the Progress of the Anabapt: Faction, by applying
yourself to the Justices to call their unlicensed School Master
to Account."[28] By a series of complicated licensing laws, it
was occasionally possible for churchmen to curtail the activities of proselytizing and unwanted sects.

Undoubtedly, the millenarianism of certain sects, such as
the Baptists led by Benjamin Keach,[29] prepared the way for
the prophets who quickly found their English converts. Names
such as Abraham Whitrow and his wife, Mr. Glover, Mary
Turner, and Elizabeth Gray appear in the lists of the chosen
visionaries.[30] Although it is nearly impossible to trace the religious backgrounds of these followers, I would hypothesize
that their religious sensibility had been cultivated by English
enthusiastic sects, or that after their experiences with the
prophets they could easily find such sects more compatible
than the established church.

The two most socially prominent supporters of the prophets
with whom we are familiar, were John Lacy, a respectable
member of Edmund Calamy's Presbyterian congregation, and
Sir Richard Bulkeley, a minor aristocrat, member of the
Royal Society, and occasional correspondent with Evelyn

[28] MSS ADD 5831, f. 149, Aug. 22, 1697, B.L.

[29] See his *Anti-Christ Stormed: or Mystery Babylon the Great
Whore, and Great City; Proved to be the Present Church of Rome*
(London, 1689).

[30] Henry Nicolson, *The Falsehood of the New Prophets Manifested
with Their Corrupt Doctrines* (London, 1708).

and the second earl of Nottingham.[31] Edmund Calamy attributed Lacy's conversion to his disaffection over the loss
of a lawsuit and claimed that Bulkeley had a hunched back
and hoped for a cure. Further evidence confirms that Bulkeley
was also extremely devout and at about the time of his conversion was chronically ill.[32] Unfortunately, there is not
enough information about the lives of either Lacy or Bulkeley
from which to piece together the story of their conversions.
Evidence does suggest, however, that Lacy at least had experienced a certain disillusionment resulting from the precarious
financial situation that existed in early eighteenth-century
England.

The prophets appear to have appealed to the economically
disillusioned and discontented. By and large their transcribed
and published prophecies are in a simple style, in language that
could have been understood by, and was expressive of, the
religious sensibility of the literate artisan segment of society.
The prophets held frequent meetings where participants were
encouraged to recount their spiritual visions and experiences;
the records of these meetings provide a fascinating account of
the fervor of both leaders and followers.[33] The activities of
the prophets aroused crowd behavior both in support of the
movement and in opposition to it.[34] In their pamphlets the
prophets and their English converts focused on the social

[31] John Evelyn, *The Diary and Correspondence* (Bohn ed., London,
1859), III, 322–323; Bulkeley to Covell, Jan. 26, 1705/6, MSS ADD
22911, f. 51, B.L.

[32] Edmund Calamy, *An Historical Account of My Own Life . . . ,*
II (London, 1829), 76–77. On the whole Calamy's account of the
prophets is highly informative. Cf. Evelyn, III, 322–323; K. Theodore
Hoppen, *The Common Scientist in the Seventeenth Century* (London, 1970), 39–40.

[33] Records can be found amid Fatio's papers in Geneva and in MSS
28.33–34, Dr. Williams's Library, some parts in shorthand of the period; and to some extent in Jean Allut et al., *Plan de la justice de Dieu
sur la terre, dans ces derniers jours, et du relevement* (London, 1714),
transcribed and edited by Fatio de Duillier.

[34] N. Luttrell, *A Brief Historical Relation of State Affairs from September, 1678 to April, 1714,* VI (Oxford, 1857), 307; Bost, ed., 163.

abuses of their time and in this manner aroused their greatest support and opposition. They preached against the power of the clergy and of lawyers. John Lacy even attacked the Whigs and low-church faction for their lack of true religious fervor.[35] One commentator described the activities of the prophets: "When they have railed against us Ministers, till they have the Rabble about them, then the levelling Principle is always taken up, and from preaching against the Priests, they turn their Doctrines against the Rich."[36]

Perhaps the most interesting and certainly the most politically radical tract to come from the prophets is Elias Marion's *Prophetical Warnings,* which couples millenarianism with an almost Blakean vision of society. The style of the tract gives us a sense of the prophet as he incited his listeners: "The time is short. The Beast is destroyed. There is no Capitulating any more. . . . Thou shalt see in a few Days the Burning of the great City. . . . 'Tis of the Beast I speak to thee; of this Harlot, of this *Babylon.* I am coming to destroy her, to deliver you from her Oppressions. . . . Tyranny shall cease among my People. There shall be nothing but Liberty. Pure Liberty without Disguise: Liberty to the Small as well as the Great."[37]

The above passage from a tract that was widely circulated, although possibly not representative of the majority of the prophets' political views, partially reveals the threat they posed as the church would have perceived it. They had coupled the millenarian vision with social protest. Once again the vision of a new heaven and a new earth had been given a

[35] See Jean Allut, *Quand vous aurez saccagé, vous serez saccagés: Car la lumière* (London, 1714); John Lacy, *Warnings of the Eternal Spirit, By the Mouth of His Servant John . . . Third and Last Part* (London, 1707), 170–171; and John Lacy, *The Ecclesiastical and Political History of Whigland of Late Years . . .* (London, 1714). Hillel Schwartz argues that Lacy is not the author of this tract.

[36] Francis Hutchinson, *A Short View of the Pretended Spirit of Prophecy, Taken from Its First Rise in the Year 1688, to Its Present State among Us* (London, 1708), 39.

[37] Marion, *Prophetical Warnings . . .* (London, 1707), 11, 13, 22.

radical definition. Marion continues, "I will overthrow the Tables of those Money Changers of the World. . . . I will spare nothing, neither King nor Princes, neither Great nor Small, all must come to an End."[38] To whomever the prophet spoke, and it appears that in these prophecies he was speaking to the poorer segments of Augustan society, he promised them a new order, an alleviation from their material concerns. He attacks the merchant class; he promises "in a few Days, to bring Damage to the Merchants of the Earth. No Trading in my Church, no cheating Merchants. I will destroy them, as an accursed thing."[39]

In attacking the merchant class, Marion is expressing sentiments that were common enough after 1688. Individuals of a religious temper, as well as others who shared a nostalgia for the older, more rigid, social order, observed often with concern the growing prosperity of the moneyed classes, and in turn feared their worldly temper. It is reasonable to assume that those who did not share in the new prosperity would have appreciated the sentiments found in Marion's writings. He concludes with an attack on the Bank of England, one of the bulwarks of this new economic order: "Yes, my Child, the Bank hath settled itself in my House. They spread about, I tell thee, from one End (of it) to the other. . . . Nothing, my Child, but Pride and Backbiting, are seen there, even in my very Palace."[40] Marion expresses throughout his writings his concern over the liaison that existed in England between the moneyed classes and the church.

Despite the present lack of precise information about the economic status of the church's membership in this period, it is a reasonably obvious assumption that the people against whom Marion preached were in part the same people who attended the Boyle lectures at St. Martin's-in-the-Fields, a parish that had one of the wealthiest congregations in London.

[38] Ibid., 64, 77.
[39] Ibid., 105.
[40] Ibid., 175.

And perhaps this assumption will explain the allusions we find in the writings of John Lacy.

Most of Lacy's early tracts, such as *Warnings of the Eternal Spirit by the Mouth of his Servant John* . . . (1707), are in the same genre as those of his fellow visionaries. In a later tract Lacy attempts to understand the opposition experienced by the prophetic movement. He postulates that *"Enthusiasm,* which exhibits as her Plea, a supernatural Influence, in Dreams, Visions, Voices heard, and forceable Impressions internal and external . . . [does] therein acknowledge the Agency of Spirits or Angels . . . and gives thereby a Testimony of the invisible World."[41] He contrasts the direct experience provided by enthusiasm to the "Philosophick Doctrines so often vented from the Pulpit, touching the secret powers in Nature." He claims that these doctrines, such as the "Corpuscular Philosophy," lead to skepticism about the existence of a supernatural inspiration behind the utterances of the prophets.[42]

Lacy appears to realize that the philosophical interests of the church provided an alternative to religious enthusiasm. The need of churchmen to explain the supernatural, and in turn to express their religious sentiment, found satisfaction in philosophical inquiry. Piety was divorced from emotion and instead wedded to the workings of the reasoning mind as it sought to comprehend both creation and creator. Enthusiasm could play no part in this rational process, and thus the activities of the prophets, especially when they coupled enthusiastic religion with even vague social protest, were dangerous and in turn prosecuted by the state.

The psychological process that led finally to the divorce of enthusiasm from religious piety began in England during the turmoils of the mid-seventeenth century. At the end of the

---

[41] Lacy, *The General Delusion of Christians, Touching the Ways of God's Revealing Himself, to, and by the Prophets* . . . (London, 1713), 496.

[42] Ibid., 497.

century this process culminated in the thinking of the churchmen who formulated the Newtonian universe. But the set of
mind necessary to produce the rational and ordered world of
the Newtonians had evolved from the troubled state of mind
of earlier philosophers such as Robert Boyle and John Wilkins.
This evolution was an extraordinarily complex phenomenon
which needs careful and thoughtful study. For the present it
must be sufficient, therefore, to emphasize that only with the
rejection of enthusiasm, expressed by the reaction of the
church to the prophets, was the formulation of a world view
as delicate and cerebral as that of the Newtonians made possible. This rejection of enthusiasm as a genuine expression of
religious piety may be seen in part as the psychological context of the Newtonian natural philosophy. In the period we
are studying, natural philosophical inquiry was a genuine
means of expression sought by men of a religious sensibility;
piety demanded thought, and the only religious emotion acceptable to churchmen was one that served as a complement
to the processes of the reasoning mind and to the workings of
an ordered society.

The church's rejection of prophetic enthusiasm rested on
one fundamental concern. As John Moore explained to the
queen in *A Sermon Preach'd . . . at St. James's Chapel,
March 8, 1705/6*, the "wild enthusiasts" threaten to destroy
the power of the civil government; for they "would found
Dominion in Grace, and . . . contend that the Servants of
the Lord of Hosts . . . are so far from owing Subjection to
any Power or Person, that they themselves only have a Right
to Rule the whole Earth." Moore compares their purposes
with those of the papists, "who having had a more Liberal
Education, and better Learning should have been more
Honest; who would vest a Power in the Vicar of Christ,
as they stile the Pope, to absolve the People of any Kingdom
from the Faith and Allegiance they bear to their Sovereigns."[43]

[43] John Moore, "A Sermon . . . March 8, 1705/6," in **Samuel
Clarke**, ed., *Sermons on Several Subjects* (London, 1715), 378.

To Moore, both enthusiasm and popery represent an equal threat to the government and by the absolutist nature of their fervor pose a threat to an ordered and balanced state. When we recall some of his sermons delivered just after the Revolution, it is evident that in Moore's pronouncements against the enthusiasts he is applying the same principles he had then formulated about the nature of government. In 1690 he preached to Queen Mary: "If therefore Men did believe that God governs the World, and ordered all the Affairs thereof, they would not ascribe the Issue of things to their own Power."[44] The providential order imposed by God on the universe and decreed for the affairs of men ensures against the tyranny of either popery or enthusiasm. In order to preserve this delicate balance of forces, religious emotion must be contained. Its expression must take a personal, almost quietistic, form that at no point interferes with the divine order inherent in civil affairs. The church therefore objected to the prophets not simply because of the threat they posed to its ecclesiastical power, but also because their religious sensibility challenged the ideology of balance and order that the latitudinarians sought to impose on social relations and market forces.

Other churchmen such as Josiah Woodward, Boyle lecturer in 1710 and a frequent visitor at the home of William Wake,[45] also used arguments similar to those we find in John Moore's sermons against the prophets.[46] Woodward was so disturbed that he wrote to Archbishop Tenison in 1709 with a long description of their activities and urged that further action be taken against them.[47] Woodward's concern is echoed in pamphlets written by John Humfrey and Offspring Blackall, Boyle lecturer in 1700. Humfrey feared that the existence of the prophets would be used by high-churchmen as an argu-

[44] Ibid., 120–121; "Of the Wisdom and Goodness of Providence, August 17 and 24, 1690," in Clarke, ed.

[45] See MSS 1770, Lambeth Palace Library.

[46] See Woodward's *Remarks on the Modern Prophets, and on Some Arguments Lately Published in Their Defense* (London, 1708).

[47] MSS 931, f. 22, Lambeth Palace Library.

ment against toleration.[48] Preaching before the queen, Blackall simply argued that there was no need at that time for any addition to revealed truth.[49]

The teachings of the prophets were especially disturbing to the churchmen known for their interest in the meaning of the Scriptural prophecies. One of the English prophets, John Lacy, admitted that he had read only Whiston on the Old Testament prophecies before he received his inspired vision.[50] Francis Misson, also a member of the prophetic movement, argued that Whiston's writings support the claim of the world's destruction within his generation.[51] Undoubtedly disturbed by the similarity found between the teachings of the prophets and his own writings, Whiston met with Lacy and attempted to convince him that evil spirits were the cause of his religious experiences.[52] Whiston relates that Humphrey Ditton, a supporter of the Newtonian natural philosophy, also attended the meeting with Lacy, and that Samuel Clarke may have held his own conference with the inspired one.[53]

According to Whiston, Clarke had read his prophetic writings and those of Joseph Mede and agreed with many of their contentions.[54] It is highly probable that Whiston's account of Clarke's interest in the prophecies is true, especially since Newton, with whom Clarke was in constant communication, shared this interest. But Clarke also shared an abiding concern for the dangerous effects wrought by enthusiastical religion. In

[48] Humfrey, *An Account of the French Prophets, and Their Pretended Inspirations* . . . (London, 1708), 45.
[49] Blackall, *The Way of Trying Prophets. A Sermon . . . at St. James's November 9, 1707* (London, 1707).
[50] John Lacy, *A Relation of the Dealings of God to his Unworthy Servant John Lacy* . . . (London, 1708), 15.
[51] Misson, *A Cry from the Desart* . . . (London, 1707), iv.
[52] W. Whiston, *Historical Memoirs of the Life of Samuel Clarke* . . . (London, 1730), 68.
[53] Ibid. Clarke was certainly familiar with Lacy's writings. See Slains 562590, April 20, 1713, Dutton-Cuninghame Correspondence, Mitchell Library, Glasgow. My thanks to Hillel Schwartz.
[54] Whiston, 156.

a 1719 letter to Archbishop William Wake, Clarke described the contents of a book he was writing and noted that in the introduction "I have added a Caution against Enthusiasts and against such as would destroy *all* external Order and Government in the Church."[55] Clarke shared with his low-church colleagues a concern for the disruptive effects produced by the overt expression of religious emotion. In *A Discourse Concerning the Connexion of the Prophecies in the Old Testament, and the Application of Them to Christ,* Clarke's interest in the prophecies extends only to their application to Christology, and he constantly emphasizes that the difficult and obscure prophecies should be studied only by individuals skilled in history.[56] The millenarian prophecies are suited for esoteric study and not for popular consumption.

In his approach to the uses of the prophecies, Clarke resembles many of his ecclesiastical peers. With the exception of William Whiston, who preached openly on the meaning of the prophecies, churchmen kept to themselves their interest in the prophetic texts. Even William Lloyd never published a full explication of his predictions derived from Scripture. Furthermore, he chided Whiston for the "extravagant fancy" found in his writings.[57] This chastisement of Whiston is ironic when we recall the precision with which Lloyd privately predicted the downfall of the beast.

No matter how fervent were the interests of the churchmen in the prophecies, after the arrival of the French prophets they increasingly avoided any discussion of their implications. The contents of the Scriptural prophecies did not simply give sup-

[55] Loose page dated May 19, 1719, in Wake MSS, CCL, XVII, Christ Church. Possibly this refers to a collection of sermons entitled *Seventeen Sermons* (London, 1724).

[56] Clarke, *A Discourse* . . . (London, 1725) 39. In this regard, Newton wrote in his *Irenicum,* "so the prophecies of both Testaments relating to Christ's second coming may remain in obscurity till that coming, and then be interpreted by divine authority." H. McLachlan, *Sir Isaac Newton Theological Manuscripts* (Liverpool, 1950), 34.

[57] Hart, 241. See also MSS ADD 24197, B.L.

port to the millenarian hopes of churchmen, to their longing for the accomplishment of the Reformation, but they served as a focus for the religious fervor of visionaries. This fervor threatened the stability of society and put into jeopardy the tenuous order so desired by churchmen as an essential prerequisite for the building of their version of the new heaven and the new earth.

After the accomplishment of the Protestant succession in 1714, that stability preached by the Newtonians increasingly became a political reality. The prophets continued their search for followers, and their activities took them to Scotland and to the Continent. They kept enthusiastic religion alive in the 1720s and 1730s, but the virulent opposition they faced from the church during the reign of Anne rendered them quietistic. Once again, on the surface at least, the Newtonian version of liberal Protestantism prevailed against the church's opposition. To complacent churchmen of the eighteenth century, and if the Methodist critique has any validity there must have been a fair number of them, the Newtonian vision of the natural world provided irrefutable justification for the public order and controlled self-interest sanctioned and maintained by church and state. The harmony of the worlds natural and political provided an intellectual and social context for the growth of empire and commercial capitalism.

The first generation of Newtonians created an ideology that permitted the achievement of that stable context. These moderates, who adapted the church to the new order created in 1688–1689, believed in the necessity of accommodating even the expression of religious feeling to the needs of an ordered society. The lesson they preached had been learned by their latitudinarian predecessors during the civil wars and Interregnum. The Newtonians took the social and scientific ideas of Restoration moderates, refined and developed them with the assistance of Newton, and applied this new ideology to the "world politick" created by the Revolution of 1688–1689. With a deep fear of the destruction that would follow the failure of their mission, the low-churchmen offered the Newtonian universe as the model for the civil and religious affairs of a nation they believed had been chosen by God to accomplish the Protestant Reformation.

Conclusion

This book argued that the Newtonians articulated their natural philosophy with constant reference to their social and political context. Thus after the Revolution of 1688–1689 Newton's philosophy of nature served as an essential support for social ideology of the liberal Anglican establishment. The main ingredients of this Newtonian ideology were stability and order in the political world, which, once instituted, would allow for the expression of "sober self-interest," even of aggression and competition, in the service of both church and state. The providence of God guided the delicate interplay of forces at work in the political world, just as he guided the natural world through active principles and laws of motion. Furthermore, the will of God, as revealed through Scripture, had decreed that the order and stability of both worlds served a historical purpose. As the first generation of latitudinarians understood it, history proclaimed the triumph of the Protestant Reformation, the destruction of antichrist, and the institution of a millenarian paradise.

For reasons that I have tried to explain in Chapter 3, millenarianism figured less and less in the thinking of Newton's followers. Nevertheless, it served in the second half of the seventeenth century as one important catalyst in making liberal Protestants into the champions and propagandizers of the new science. Among those younger churchmen, such as Bentley and Clarke, who thought more about this world than about the new one yet to be instituted, their social ideology, in essence very similar to that of their latitudinarian predecessors, was intended primarily to advance the interests of the Anglican church and to ensure its well-being in the social and political context sanctioned by the Revolution of 1688–1689.

In short, the meaning of the Newtonian synthesis that was bequeathed to the Enlightenment, partly by Newton himself, was profoundly political. The Newtonian ideology was intended to promote liberal Protestant interests at home and abroad and to maintain and bolster the power and status of a pious and God-fearing elite. All demands for reform, if they came from political radicals like the freethinkers or from religious enthusiasts like the French prophets and their followers, were regarded by the Newtonians as dangerous because they were seen as threatening to decrease the church's power or to alter the existing fabric of social order and political stability.

If my analysis of the nature and intention of the Newtonian ideology, as it was initially formulated, appears credible to students of the Enlightenment it will be necessary to rethink the role of Newtonianism in eighteenth-century thought, and also to take a new look at the freethinking tradition that begins and ends, roughly, with Toland and d'Holbach. But the implications of this study are a subject for the future. For the present I shall be content to have shown that Newton's philosophy of nature developed within a social context and served as the underpinning of a social and political ideology that attempted to ensure stability and piety and yet allow for the expression of individual self-interest, which in the course of the eighteenth century became increasingly synonymous with capitalistic enterprise.

The Boyle Lectures
1692–1714

1692 Bentley, R. *The Folly and Unreasonableness of Atheism.* London, 1692.

1694 Kidder, R. *A Demonstration of the Messias.* 3 vols. London, 1694, 1699, 1700.

1695 Williams, J. *The Possibility, Expediency and Necessity of Divine Revelation.* London, 1695–1696.

1696 Williams, J. *The Perfection of the Evangelical Revelation.* London, 1696.

1697 Gastrell, F. *The Certainty and Necessity of Religion in General.* London, 1697.

1698 Harris, J. *Immorality and Pride, the Great Causes of Atheism.* London, 1698.

1699 Bradford, S. *The Credibility of Revelation, from Its Intrinsic Evidence.* London, 1700.

1700 Blackall, O. *The Sufficiency of a Standing Revelation.* London, 1717.

1701–2 Stanhope, G. *The Truth and Excellence of the Christian Religion Asserted against Jews, Infidels, and Heretics.* London, 1702.

1703 Adams,   . [Not printed.]

1704 Clarke, S. *Demonstration of the Being and Attributes of God.* London, 1706.

1705 Clarke, S. *A Discourse Concerning the Unchangeable Obligations of Natural Religion, and the Truth and Certainty of the Christian Revelation.* London, 1706.

1706 Hancock, J. *Arguments to Prove the Being of God.* London, 1707.

1707   Whiston, W. *The Accomplishment of Scripture Proph-ecies.* Cambridge, 1708.
1708   Turner, J. *The Wisdom of God in the Redemption of Man.* London, 1709.
1709   Butler, L. *Faith and Practice of True Christians No Matter of Shame or Reproach.* London, 1711.
1710   Woodward, J. *The Divine, Original, and Incomparable Excellency of the Christian Religion, as Founded on the Holy Scriptures, Asserted and Vindicated.* London, 1712.
1711–12   Derham, W. *Physico-Theology: or, A Demonstration of the Being and Attributes of God, from the Works of Creation.* London, 1713.
1713–14   Ibbot, B. *On the Exercise of Private Judgement, or Free-thinking.* London, 1727.

Bibliographical Note

Since I have tried to list all pertinent material in the footnotes, there seems no necessity to reprint these sources in a complete bibliography. Instead I cite here only important primary and secondary sources which were crucial to the development of this book.

Almost every subject with which this essay has dealt required the discovery of manuscript material for its satisfactory explication. It is impossible to overemphasize the importance of archival material for uncovering private connections, thoughts, or hostilities, all of which must be known if we are to understand statements made by public men. Newton's manuscripts in the Cambridge University Library and at King's College, especially ADD 3970, U.L.C., have been especially valuable. The University Library provided much important information: Simon Patrick's manuscripts, John Moore's unpublished sermons, MSS Dd. 14.9, Dd. 14.15, the Allix MSS, and ADD 7647 with reference to Newton. Thomas Burnet's own copy of the *Sacred Theory* deposited there was also essential. The collections of the British Library are rich and yielded unexpected treasures: ADD 4236 on Halley and Tillotson; ADD 10039, f. 63, Burnet to Southwell; ADD 24197, Lloyd and Whiston; Lansdowne 1024 on Boyle's will; ADD 28104 on Heneage Finch's involvement in the Boyle lectures; ADD 29573 on the circulation of Socinian pamphlets; and ADD 4295 and 4465, Toland's manuscripts. For the freethinkers the Shaftesbury manuscripts at the Public Record Office must be consulted, and the Marchand papers at the University Library, Leiden, The Netherlands, are yet to be explored fully.

In Oxford, Evelyn's papers at Christ Church Library are full of fascinating material on social and intellectual life from the 1650s onward. At the Bodleian the Tanner MSS are rich, and Locke's correspondence, which I hope will soon see publication, should always be consulted by students of the 1690s.

Lambeth Palace Library, in particular MSS 1770 and 933, is important for the interests of latitudinarian churchmen. At Dr. Williams's Library, London, there are useful manuscript sources for William Lloyd and Fatio de Duillier. Most of Fatio's papers are in Geneva at the Bibliothèque Publique et Universitaire. The Southwell letters in the Forster collection of the Victoria and Albert Museum Library added a necessary piece to the puzzle surrounding Burnet's *Sacred Theory*. Finally, the Gregory MSS at the Royal Society contained information on the French prophets. The fact that material turned up often in unexpected places, for example, the Gregory papers, confirmed the usefulness of an interdisciplinary approach to these topics.

Certain obvious published sources must be read and interpreted. All the sermons of the latitudinarians, Barrow, Tillotson, Moore, Wilkins, Tenison, Patrick, and Fowler, have been read and the relevant texts cited in the footnotes. Certain diaries and memoirs proved essential: E. de Beer, ed., *The Diary of John Evelyn* (London, 1955); William Whiston, *Historical Memoirs of the Life of Dr. Samuel Clarke* (London, 1730); W. G. Hiscock, *David Gregory, Isaac Newton and Their Circle* (Oxford, 1937); Thomas Hearne, *Remarks and Collections* (Oxford, 1885–1921); Alexander Gordon, *Freedom after Ejection, A Review (1690–92) of Presbyterian and Congregational Nonconformity in England and Wales* (Manchester, 1917). Certain key letters are available in H. W. Turnbull, ed., *The Correspondence of Isaac Newton*, 4 vols. (Cambridge, 1959–1967); and on Toland in Jacques de Chaufepié, ed., *Nouveau dictionnaire historique et critique . . . de Mr P. Bayle* (Amsterdam, 1750), and in G. Bonno, ed., "Lettres inédites de LeClerc à Locke," *University of California Publications in Modern Philology*, 52 (1959).

But these primary sources, in order to be understood fully, had to be seen in the light of what other scholars have said about the late seventeenth century. Certain books have made indelible impressions on my thinking and must be singled out for mention. For the civil wars and Restoration ideological background there is Christopher Hill's *The World Turned Upside Down* (London, 1972); C. B. Macpherson's *The Political Theory of Possessive Individualism, Hobbes to Locke* (London, 1962); R. W. Tawney, *Religion and the Rise of Capitalism* (New York, 1961, reprint), and most important, Richard Schlatter, *The Social Ideas of Religious Leaders, 1660–1688* (Oxford, 1940). Also useful is Barbara Shapiro, *John Wilkins, 1614–1672. An Intellectual Biography* (Berkeley, 1969), and P. Rattansi, "Paracelsus and the Puritan Revolution," *Ambix*, 11 (1964), 24–32. A book with which I have been familiar since its inception is invaluable: J. R. Jacob, "Robert Boyle and the English Revolution," to be published by Burt Franklin, Inc.

In order to relate ideas to their social context in this later period I benefited enormously from G. Straka, *The Anglican Reaction to the Revolution of 1688* (Madison, Wis., 1962); J. H. Plumb, *The Growth of Political Stability in England, 1675–1725* (London, 1967); Geoffrey Holmes, *British Politics in the Age of Anne* (London, 1967); and H. G. Horwitz, *Revolution Politicks: The Career of Daniel Finch, Second Earl of Nottingham, 1647–1730* (Cambridge, 1968).

For bibliography on Newton there are many items discussed in Chapter 5, but I would single out J. E. McGuire and P. Rattansi, "Newton and the Pipes of Pan," *Notes and Records of the Royal Society*, 21 (1966), 108–143; D. Kubrin, "Newton and the Cyclical Cosmos: Providence and the Mechanical Philosophy," *Journal of the History of Ideas*, 28 (1967), 325–346; Christopher Hill, "Newton and His Society," in Robert Palter, ed., *The "Annus Mirabilis" of Sir Isaac Newton, 1666–1966* (Cambridge, Mass., 1970); Frank Manuel, *A Portrait of Isaac Newton* (Cambridge, Mass., 1968).

For church history two items in particular provided necessary information: G. V. Bennett, "King William III and the Episcopate," in *Essay in Modern English Church History*, eds. G. V. Bennett and J. Walsh (London, 1966); and A. Tindal Hart, *William Lloyd, 1627–1717* (London, 1952). Many other items are cited in the text, but it should be noted that church historians tend to ignore intellectual life and concentrate instead on ecclesiastical structure and politics.

The freethinkers deserve a separate study. I have been con-

cerned with them since 1965 and at that time the book that revolutionized my thinking was Frances Yates, *Giordano Bruno and
the Hermetic Tradition* (London, 1964). But the revolution did
not stop there; I can only recommend her *The Art of Memory*
(London, 1966), and *The Rosicrucian Enlightenment* (London,
1972). All who work in this field have benefited from the contributions of Rosalie Colie, in particular, "Spinoza in England,
1665–1730," *Proceedings of the American Philosophical Society*,
107 (1963). For the importance of the republican freethinkers for
the Enlightenment see Franco Venturi, *Utopia and Reform in the
Enlightenment* (Cambridge, 1971).

Finally, if a student were coming to this period for the first
time and looking for a useful introduction, I can still recommend
Paul Hazard, *The European Mind, 1680–1715* (New York, 1963);
and Hélène Metzger, *Attraction universelle et religion naturelle
chez quelques commentateurs anglais de Newton* (Paris, 1938).

Index

(Compiled with the assistance of Margaret T. Fidanza)